Subverting Consumerism
Reuse in an Accelerated World

W0113696

There is now a widespread interest in reuse in many domains, from opera houses built over old warehouses, to vintage clothes and everyday goods incorporating repurposed materials or parts. Despite its ubiquity, this extensive creative work is typically seen in narrowly environmental terms, as a means of reducing carbon, resource use or waste. However, as this volume shows, reuse also has aesthetic and cultural dimensions and a rich social currency, invoked to consciously subvert the accelerated consumer culture responsible for our unfolding environmental crisis.

In three parts, the essays in this book consider reuse in terms of values, aesthetics and meaning, its application in contemporary urban and spatial settings, and the revival of social practices involving a more conscious recourse to reuse and repair. These are bookended by the editors' essays: the first, on the significant relationship between reuse and technological and social acceleration evident in the surrounding consumer society; and the last, on the multiple forms of reuse deployed in a contemporary alternative building practice, and their contributions to presenting alternative ways of living in the world.

Challenging dominant understandings of 'waste' and 'consumption', *Subverting Consumerism* shows how reuse has become a means for many to creatively engage with the past, and to discover a continuity and sense of place eroded by the accelerative regimes of contemporary consumerism. Becoming a means of resistance, and offering a range of aesthetic, social and economic possibilities, reuse can be found to subvert and challenge the obsessive quest for the new found in contemporary consumerism.

Robert Crocker (DPhil.) teaches the history and theory of design and design for sustainability in the School of Art, Architecture and Design at the University of South Australia, where he is Deputy Director of the China Australia Centre for Sustainable Urban Development. His research is focused on consumption and its contributing role in our environmental crisis. His most recent book is *Somebody Else's Problem: Consumerism, Sustainability and Design* (Greenleaf/Routledge 2016).

Keri Chiveralls (PhD.) is the Discipline Lead and Head of Course for the first full degree postgraduate courses offered in Permaculture Design at CQUniversity. She teaches in permaculture and related topics and her research interests are in cultural and environmental anthropology, social movement studies and theories of social change. She received her doctorate in Anthropology/Social Inquiry at the University of Adelaide in 2008. This is her first co-edited book publication.

Antinomies
Innovations in the Humanities, Social Sciences and Creative Arts

Series Editors: Anthony Elliott and Jennifer Rutherford
Hawke Research Institute, University of South Australia

This series addresses the importance of innovative contemporary, comparative and conceptual research on the cultural and institutional contradictions of our times and our lives in these times. *Antinomies* publishes theoretically innovative work that critically examines the ways in which social, cultural, political and aesthetic change is rendered visible in the global age, and that is attentive to novel contradictions arising from global transformations. Books in the Series are from authors both well-established and early careers researchers. Authors will be recruited from many, diverse countries – but a particular feature of the Series will be its strong focus on research from Asia and Australasia. The Series addresses the diverse signatures of contemporary global contradictions, and as such seeks to promote novel transdisciplinary understandings in the humanities, social sciences and creative arts.

The Series Editors are especially interested in publishing books in the following areas that fit with the broad remit of the series:

- New architectures of subjectivity
- Cultural sociology
- Reinvention of cities and urban transformations
- Digital life and the post-human
- Emerging forms of global creative practice
- Culture and the aesthetic

Culture, Identity and Intense Performativity
Being in the Zone
Edited by Tim Jordan, Brigid McClure and Kath Woodward

Subverting Consumerism
Reuse in an Accelerated World
Edited by Robert Crocker and Keri Chiveralls

For a full list of titles in this series, please visit www.routledge.com/Antinomies/book-series/ANTIMN.

Subverting Consumerism

Reuse in an Accelerated World

**Edited by Robert Crocker
and Keri Chiveralls**

Routledge
Taylor & Francis Group

LONDON AND NEW YORK

First published 2018
by Routledge
2 Park Square, Milton Park, Abingdon, Oxon OX14 4RN

and by Routledge
52 Vanderbilt Avenue, New York, NY 10017

First issued in paperback 2020

Routledge is an imprint of the Taylor & Francis Group, an informa business

British Library Cataloguing-in-Publication Data
A catalogue record for this book is available from the British Library

Library of Congress Cataloging-in-Publication Data
Names: Crocker, Robert, 1952– editor. | Chiveralls, Keri, 1979– editor.
Title: Subverting consumerism : reuse in an accelerated world / edited by Robert Crocker and Keri Chiveralls.
Description: Abingdon, Oxon ; New York, NY : Routledge, 2018.
Identifiers: LCCN 2018006777 | ISBN 9781138189096 (hardcover) | ISBN 9781315641812 (e-book)
Subjects: LCSH: Consumption (Economics)—Social aspects. | Environmentalism. | Sustainable urban development. | Recycling (Waste, etc.)—Social aspects.
Classification: LCC HC79.C6 S83 2018 | DDC 306.3—dc23
LC record available at https://lccn.loc.gov/2018006777

ISBN 13: 978-0-367-66595-1 (pbk)
ISBN 13: 978-1-138-18909-6 (hbk)

Typeset in Times New Roman
by Apex CoVantage, LLC

Contents

Acknowledgments

There are many people we need to thank here. Firstly, we greatly appreciate the contribution of the authors of the essays included in this book, and would like to thank them for their hard work, commitment to the project, and patience dealing with the review and editing process. We would also like to thank those many friends and colleagues who at various stages became involved in the development of this book, including those who anonymously reviewed these essays, and those who read them and provided additional comments as the project developed. We owe a special thanks to Desirée Bernhardt for her excellent and thoughtful editing of one complete draft of the book. We would also like to acknowledge and thank Stuart Walker and Jonathan Chapman for their initial comments on our proposal for the book. And we would also like to thank our friends at Routledge, including our editors, Alyson Claffey and Diana Ciobotea, and of course Anthony Elliott, our very patient book series editor.

It seems important here to also acknowledge the inspirational role of the Unmaking Waste conference held at the University of South Australia in May 2015, where some earlier versions of these chapters were first presented, in particular those by Robert Crocker, Hing-Wah Chau, Gini Lee, Kim Fraser, and Keri Chiveralls. The idea for this book really emerged from the experience of this conference, and the excellent exhibition that a number of the presenters, including Stuart Walker, contributed to. We should acknowledge here especially the efforts of Kate Thornton and Chris Thornton (no relation!) for their extremely hard work making this conference the memorable experience it became for everyone, and Gini Lee for her own work on curating this exhibition.

Finally, last but not least, we would like to thank our families for their patience and support over the last few years, Merrie Russell, Rose Crocker and Lakshmi Crocker, and Olive, Ian and Emma Blackman, Keith and Judy Chiveralls, Sarah, Steve, Freya and Owen Finney, as well as Joan and Mike O'Keefe.

Illustrations

Contributors

Linda Blackall is a Professor of Environmental Microbiology at the University of Melbourne; she is also an Adjunct Professor at Swinburne University of Technology, Melbourne. Her research is in complex microbial communities, and she has published over 150 peer-reviewed journal articles in the fields of biological wastewater treatment, marine and coral microbiology, mammalian microbiomes, and molecular method development in microbial ecology.

Sally Butler is Associate Professor in Art History at the University of Queensland where she teaches and researches in the areas of global visual literacy, visual politics, cross-cultural aesthetics, and Australia Pacific indigenous art. She publishes widely in visual arts magazines and has published four books on Australian art.

Hing-wah Chau is the Melbourne Early Career Academic Fellow in Architectural Design and the Senior Tutor of Masters Design Studios at the University of Melbourne. He has been teaching architectural design at the University of Melbourne since 2011. He has more than ten years of professional practice experience in Hong Kong as a registered architect, working in an interdisciplinary environment and involved in different types and scales of projects.

Keri Chiveralls (PhD.) is the Discipline Lead and Head of Course for the first full degree postgraduate courses offered in Permaculture Design at CQUniversity. She teaches in permaculture and related topics and her research interests are in cultural and environmental anthropology, social movement studies and theories of social change. She received her doctorate in Anthropology/Social Inquiry at the University of Adelaide in 2008. This is her first co-edited book publication.

Tim Cooper is Professor of Sustainable Design and Consumption at Nottingham Trent University, where he specializes in research into product lifetimes. He is Head of the Sustainable Consumption and Clothing Sustainability research groups and Co-Director of the Centre for Industrial Energy, Materials and Products (CIE-MAP). In 2015 he established the biennial PLATE (Product Lifetimes and the Environment) conferences. He has written widely on sustainable consumption and design, and was contributing editor of *Longer Lasting Products* (Routledge, 2010).

Robert Crocker (DPhil.) teaches the history and theory of design and design for sustainability in the School of Art, Architecture and Design at the University of South Australia, where he is Deputy Director of the China Australia Centre for Sustainable Urban Development. His research is focused on consumption and its contributing role in our environmental crisis. His most recent book is *Somebody Else's Problem: Consumerism, Sustainability and Design* (Greenleaf/Routledge 2016).

Kim Fraser is a fashion designer and academic in Aotearoa, New Zealand. Her practice-led research is grounded in socially responsible design and has resulted in several co-authored publications concerning sustainable fashion practices and post-consumer textiles waste. She is currently working on projects that promote strategies for more sustainable consumption and production of fashion textile product.

Claire Langsford recently completed her PhD in Anthropology in the School of Social Sciences at the University of Adelaide. Her research on the material culture of fan communities has been published in *Sites: a journal of social anthropology and cultural studies* and in *Manga Vision: Cultural and Communicative Perspectives* (Monash University Press 2016). She is currently working with COTA SA (Council of the Ageing) on a project exploring older people's lived experiences of built and natural environments.

Gini Lee is the Elisabeth Murdoch Chair of Landscape Architecture at the University of Melbourne. Her research investigates ways in which designed landscapes are incorporated into the cultural understandings of individuals and communities. Her recent landscape curation practice is an experiment with Deep Mapping methods to investigate the water landscapes of remote and rural Australia. She is a registered landscape architect.

Max Liboiron is a scholar, artist, and activist whose work focuses on how invisible, harmful, emerging phenomena such as 'slow' disasters and toxicants from plastics become apparent in science and activism, and how these methods of representation relate to action. Her social/science lab, Civic Laboratory, creates action-oriented research through grassroots environmental monitoring, mainly of marine plastics. She also runs the widely respected *Discard Studies* blog, an interdisciplinary online hub for research on waste.

Peter Newton is a Research Professor in Sustainable Urbanism in the Centre for Urban Transitions at the Swinburne University of Technology in Melbourne. He has led many research projects on sustainable consumption, green economy transitions, and urban regeneration. He is currently Research Leader in the multi-institution Cooperative Research Centre for Low Carbon Living.

Giuseppe Salvia is a Research Fellow at Politecnico di Milano in Italy. He has been engaged in national- and EU-funded projects on sustainable practices, grassroots innovation socio-technical transformations. He is particularly interested in the relationship between people and products in order to promote a shift towards sustainable patterns of consumption and production through design.

Cathy Smith is an Australian architect, interior designer and academic. She is the inaugural Turnbull Foundation Women in the Built Environment scholar at the University of New South Wales (2018–2020); and a recipient of a Richard Rogers fellowship and residency, Harvard University Graduate School of Design (Fall 2018), focused on property guardianship in London. She has taught design, history and theory, and construction at the University of Newcastle, the University of Queensland and the Queensland University of Technology. Her research on DIY production has been published in Australian Feminist Studies, Architectural Histories, Interstices, Architectural Theory Review, IDEA and Design Ecologies.

Vivienne Waller is a Senior Lecturer in Sociology at the Swinburne University of Technology in Melbourne. Her research interests are mainly in environmental practices and knowledge, food waste management, and the sociology of the environment. She has published in the areas of regulatory policy, information systems, and technology use.

Introduction

Robert Crocker and Keri Chiveralls

> *Our enormously productive economy demands that we make consumption our way of life, that we convert the buying and use of goods into rituals, that we seek our spiritual satisfactions, our ego satisfactions, in consumption. The measure of social status, of social acceptance, of prestige, is now to be found in our consumptive patterns. The very meaning and significance of our lives today are expressed in consumptive terms . . . . We need things consumed, burned up, worn out, replaced, and discarded at an ever-increasing pace.*
>
> *(Lebow 1955, p. 7)*

Frequently circulated on Facebook with titles like "Consumerism" or "Excess" above it, this quotation from a 1955 essay by the US economist and marketing expert Victor Lebow is suggestive of how consumerism is now seen as not only an individual problem of excess, but also as the result of a systemic manipulation by manufacturers and retailers of consumers and their desires. It is a popular quotation perhaps also because it says what many instinctively feel: that excessive accumulation is part of a 'system' seemingly stacked against them, a system 'designed' to encourage them to buy more than they might need, and to waste more than should be wasted (Urry 2010; Lipovetsky 2011).

Within this system, to design and manufacture the new is to design and produce what will be acquired, accumulated, and then discarded, more rapidly to be replaced by the new (Slade 2006; Smart 2010). As Lebow's words suggest, it is the increasing efficiencies of this "enormously productive" increasingly global system dedicated to production for consumption, with its continuous technical advances and improvements in every area, that soon drives real prices down and makes more things available to more people more quickly, a now global phenomenon that Lipovetsky (2011) aptly names "hyper-consumption."

Persuading more people to acquire new products all year round ensures that the factories will continue to produce, and the economy will expand (Dinnin 2009). At the time Lebow was writing, it was widely believed that making more consumer goods for more people would raise the living standards of everyone within that nation (Lipsitz 1998; Cohen 2004), a belief that has attained sanctity today as mainstream economic policy, despite the advent of globalization, despite the lowering of real prices relative to most incomes in 'developed' or high consuming nations (Schor 2005; Cohen 2011), and despite the increasingly high environmental costs of the resulting hyper-consumption (Pfister 2010; McNeill and Engelke 2014). This linear economy of continuous growth, of 'make, use, and waste,' inevitably relies on shorter retention rates and more frequent disposal to grow (Wieser 2016), for without the frequent replacement of what most need, there is simply no room in their lives for the number and variety of new things they might be persuaded to buy (Liboiron 2013).

Consumerism is the result of this accelerated economy of 'make, use, and waste,' and is typically defined as the now increasingly universal "state of mind and way of life" associated with the buying or possessing of goods and services this hyper-consumption necessarily involves (Smart 2010, pp. 8–10, and Crocker, this volume). While consumption can be a vehicle for self-expression, love, and the fostering of relationships (Miller 1998), comparing what is possessed with what others have, or what is being offered for sale, can drive the expected standards associated with consumption ever higher, encouraging individuals to want, 'need', and strive for more, propelled by the promise of the new (Dittmar 2007; Dinnin 2009; Tatzel 2014).

It is thus no accident that in this time of hyper-consumption, consumerism, and environmental crisis, there is now a widespread and increasingly visible reassessment of reuse and its potential application from many different disciplines and approaches (Alexander and Reno 2012; Cooper and Gutowski 2015; Aytac, Arslan, and Durak 2016). Reuse is of more interest to designers and architects now than it has been for most of the twentieth century, and also to state and city authorities struggling to reduce ever-increasing flows of wastes (UNEP and ISWA 2015). Reusing components or parts, and remanufacturing products, is also of increasing interest to larger manufacturers concerned with increasing resource scarcity (Grantham 2011; Cooper and Gutowski 2015). Reusing older materials, or including parts of older things in newer ones, has become increasingly common and is apparent in fashion, furniture, furnishing, and design, with many lifestyle magazines now showing tastefully recovered goods alongside new ones, often from traders specializing in reclaimed and 'vintage' goods (Brace-Govan and Binay 2010). Some have more radically adopted 'alternative' consumption practices involving more forms of reuse, including sharing or swapping everyday services or goods as a means of reducing a reliance on more new 'stuff' (Soper 2007; Cherrier 2010; Evans 2011).

Reuse today is thus a complicated phenomenon, since it includes multiple variations and spans or cuts through different domains and groups of users (Appelgren and Bohlin 2015; Cooper and Gutowski 2015). Reuse now includes more than just the recovery and circulation of second hand goods, but also the deliberate recovery and reuse of what was once regarded as unusable and unsalable, including the 'upcycling' of unmarketable wasted products such as food, broken metals, timber, or e-wastes, and even toxic chemicals (Cooper and Gutowski 2015; Velis 2014). In many domains reuse has become an increasingly visible practice, both within and outside the more formal economies of monetary exchange, and the technical worlds of 'resource recovery', waste management, material recycling, and industrial reuse, and also of the professions of 'heritage' preservation and restoration in design (Appelgren and Bohlin 2015). Newer, emerging reuse-oriented practices, often linked to notions of a 'circular economy' (Webster 2015; Velis 2014), share some significant characteristics that help us distinguish these from the more traditional practices of reuse, once commonly employed in the past (Strasser 1999; 2003).

Firstly, as noted above, reuse as a multitude of varied practices and activities at a variety of scales has grown remarkably, alongside, and often within, growing global and local markets for new goods and 'secondary' markets of used and waste goods, now larger in number and variety than they have ever been before (Gregson and Crewe 2003; Appelgren and Bohlin 2015; Cooper and Gutowski 2015). For as new goods multiply, and as selling the new alongside the used becomes more common, enabled particularly by social media and more formal platforms like e-Bay and Amazon, reuse has become an alternative, affordable pathway to consumption for more people, with some significant social and economic consequences (Ritzer et al. 2012; Gregson et al. 2013).

Secondly, many practices involving reuse now involve a more active engagement in, or at least a renewed appreciation of, a large variety of older, once neglected or marginalized craft skills, and a renewed willingness to DIY (do it yourself) or DIT (do it together), often aided by new technologies of information and making, to repair, remake, or upcycle older 'found' goods in many domains (Watson and Shove 2008; Holroyd et al. 2015; and see Langsford, Smith, and Chiveralls, this volume). This interest has been stimulated by online media like YouTube, which encourages fellow travelers in many domains to exchange their knowledge, learn new skills, show off their work, and exchange or sell it (Luckman 2013; 2015). Demonstrations of how to make and play an instrument, draw, knit, crochet or embroider, throw a pot, restore or make furniture, build a tiny house, or learn other now neglected craft skills can all be found online, with viewers in their thousands, sometimes hundreds of thousands, watching and learning (Luckman 2015; Holroyd et al. 2015).

Thirdly, as a growing fascination with reuse and handcraft has gathered pace, skills in reuse and restoration have spread into mainstream trades, design, and architecture practices (Baum and Christiaanse 2012). Many large urban architectural projects, for example, can now involve substantial adaptive reuse projects, perhaps of a former industrial or commercial building or structure (Ward 2012; Young 2012, and Crocker, this volume), while many interior design projects now involve the upcycling of waste timbers, steels, papers, textiles, or other found materials, or the commissioning of art works from recycled materials (Baum and Christiaanse 2012; Lee, Chau, and Chiveralls, this volume). Herzog and de Meuron's stunning new, billion-dollar Elbphilharmonie concert hall in Hamburg, for example, is one of several new iconic buildings constructed on top of, or making use of, an older structure, in this case an abandoned brick factory (Kennicott 2017; and see Crocker, Chau, and Lee, this volume).

Fourthly, this rise of reuse has been accompanied by a revival in all kinds of DIY and professional repair, with repair cafés springing up in many places and groups dedicated to repairing everything from toasters to mobile phones and computers attracting considerable interest (Gregson, Metcalfe and Crewe 2009; Ritzer et al. 2012; Gregson et al. 2013; and see Cooper and Salvia, and Chiveralls, this volume). Many bicycle shops, for example, will now not only service and repair older bicycles, but also restore or upcycle for resale more fashionable but older wrecked machines that once would have been sent to the recycler's. Some larger companies are also now offering more circular repair services, with the European Union, Japan, and South Korea, amongst others, changing laws and regulations to encourage repair's revival to help reduce waste flows and increase post-industrial employment options (see Cooper and Salvia, this volume). Many studies indicate that repair and reuse are more economically viable than they have been before, largely due to new information, design, and manufacturing technologies (e.g. Duflou, Van, and Dewulf 2013, Cooper and Gutowski 2015).

Finally, there are signs that this increasing mainstreaming of reuse is influencing consumption itself and how the new itself is perceived as a category, both in terms of its inherent value and of its identity, 'biography', or narrative (Gregson et al. 2013; Appelgren and Bohlin 2015; and see Crocker, Liboiron, Lee, and Chiveralls, this volume). The reused – whether object, building, landscape, or service – is seen and valued by its users in a manner distinct from the way the same buyers might regard a new product or building: the value of the reused object, image, or space is often tied to a more personal narrative of prior use, abandonment, discovery, or restoration (Appelgren and Bohlin 2015, and Chiveralls, this volume).

In this volume of essays, we have tried to focus attention on reuse in terms of design and creative practice, in addition to emphasizing reuse's significant social, cultural, and environmental impacts. The dominant idea behind this book is that the world of new products and buildings made entirely from virgin materials is headed for a rapid decline (Smart 2010; Grantham 2011; Urry 2010; Webster 2015). The world is now full of the new and of expanding volumes of detritus created by the once new, or so it seems (Bourriaud 2002): scarce resources, enduring economic inequalities, changing social and political expectations, tighter regulations, and more demanding environmental problems (Dauvergne 2008; 2010; Graeber 2012), are all conspiring to push design and other creative practice into working with and through reuse. This may well be one of the most significant moments in the development of design and other creative fields for half a century or more, and one directly related to, and interacting with, our unfolding environmental crisis (Fry 2009; Walker 2011).

Many of the essays in this book are suggestive of this transformation, and where we might look to find the beginnings of such a change and how this change is playing out in specific cases or examples. For reuse does not only involve manipulating a devalued, abandoned, or wasted product or environment for some unique or valued material effect, but a different way of seeing the world and its parts, where the already existing may appear 'good enough', and as sufficient in itself, perhaps with some slight modification. This might include an insertion of the old into the new, or a rearrangement of older parts and perhaps their realignment into a new set of relations (Baum and Christiaanse 2012, and Chiveralls, this volume). As this suggests, reuse encompasses not just the conversion of one material into another more useful, but also more broadly an understanding of how these materials might be handled, related to, transformed, remade, or repaired, joined together with the new, and how they might come to be (re)valued, sold, and adopted for use (Graeber 2012; Alexander and Reno 2012).

We have divided the book, somewhat loosely, into three parts, bookended by our own contributions as editors. The first, by Robert Crocker (Chapter 1), is an opening discussion of how reuse has become more significant within a period of unprecedented and rapid technological and social change or 'acceleration'. By drawing attention to how strategies of 'deceleration' or 'slowing down' are often deployed during periods of intense acceleration, this essay suggests that this push towards reuse is in part an expression of deeper longings for continuity and place, for connection and relationship, for human-material redemption in a time of multiple discontinuities, unprecedented acceleration, and environmental crisis (see Graeber 2012). The last essay in the book, by Keri Chiveralls (Chapter 11), takes as its starting point the practice of building an 'Earthship' – an earth-bermed house made from 'waste' and other materials, including tires, bottles, and cans. The chapter explores how the building of these structures remakes the builders' relationship with place, with the materials they are using, and with one another. Such reuse can be understood as a deliberative project of value transformation that challenges dominant paradigms and cultural constructions while building alternative social and physical structures from the 'ruins of modernity' (Dawdy 2010).

The first part of the book, which follows Robert Crocker's opening essay, is largely concerned with reuse from the perspective of "Culture, Meaning, and Value." It considers the many practices and diverse economies (Gibson-Graham 2008) involved in reuse and their social and cultural implications, a theme that Keri Chiveralls's final essay returns to at the end of the book. The first essay in this section by Max Liboiron (Chapter 2) is an investigation of how and why we come to value our possessions, beyond the simple monetary exchanges we engage with in everyday life. Using a gallery-based experiment involving the exchange and gifting of carefully presented and selected 'waste' products, Liboiron asked her gallery's visitors to exchange what they liked most among the exhibits for something of seemingly equivalent value, which could be anything they value from their own lives. Cutting through some of the traditional distinctions between exchange and gifting, between waste and art, between monetary worth and value, this essay shows how alternative or 'diverse economies' of value can be seen to operate within and alongside everyday monetary exchanges and relationships.

The next essay, by Sally Butler (Chapter 3), looks at how museums, aware of how little of their often-vast collections are regularly shown, have come to more actively engage with specific groups of users to rediscover and display buried or neglected parts of their own collections. In her essay, she traces how one group of Torres Strait Islander artists from northern Australia has been making art informed by the artwork of ancestors housed in an older collection in the museum. This has preserved long lost cultural artifacts from the group's ancestors. Making use of this collection to expand their understanding of their own history, these indigenous artists in turn work their own art into an intriguing dialogue with what they have seen in the museum.

In the last essay in this part, Claire Langsford (Chapter 4) reflects on another type of reuse, that is the making of imaginative costumes by a group of 'cosplayers'. These cosplayers creatively reuse waste materials, including card, plastic, and textiles, to dress up for the parts their in-costume role-playing demands, swapping and sharing not only their special creations with each other, but also their skills in designing and making these often fragile, elaborate, and short-lived costumes. Here again, questions of value and meaning, exchange, material circulation, and gifting are raised, with the value read into these costumes by cosplayers moving in and out of the standard economic categories typically imposed on the world of goods. As in the previous essays, the role of creative skill in producing value in each practice is raised, as is the meaning of what is being produced and consumed.

The second part of the book, "Strategies and Landscapes of Reuse," is more directly concerned with the world of design and the various approaches designers now use to engage with reuse, particularly with the forms of reuse now being employed in architecture, landscape, and urban design. This begins with Cathy Smith's essay (Chapter 5) on a project to temporarily reuse abandoned commercial buildings and shop-fronts in Newcastle, Australia, a city experiencing the kind of post-industrial malaise now typically found all over the industrialized world, from Detroit to Birkenhead. In this essay, she is particularly concerned to highlight how reuse is drawn upon, sometimes with a measure of success, as a strategy for reinserting life and economic activity into urban spaces that have been abandoned or destroyed by globalization, industrial decline, and neoliberal economic policies.

This is followed by an essay by Hing-Wah Chau (Chapter 6) on the architecture of the Prizker prize-winning Chinese architect Wang Shu, who often works with old building remnants such as traditional clay roof tiles and bricks, inserting them into his large-scale civic buildings, precincts, and installations. Drawing public attention to the rapid destruction of traditional buildings and cityscapes in the name of 'development' in China, Wang's architecture works not just as a political statement but as a form of historical and cultural remediation, reminding us of continuities that have been overwhelmed or driven to the margins by the sheer scale and speed of China's extensive industrial and urban development.

The final essay in this section, by the landscape architect Gini Lee (Chapter 7), draws attention to the possibilities of 'making do' in spatial design, taking what already exists in an often violent and unequal material world and making smaller, adaptive changes to achieve what might be required. In this essay, three cases are explored to show how designers can work with very unpromising materials and limited budgets to transform and restore human interactions and relations. Such practices have the potential to reduce the effects of conflict, natural disaster, social inequalities, and dispossession. Like the preceding chapters in this section, Lee's demonstrates how design strategies can work with, rather than against, what remains, and can revalue and redeem what has been wasted, deprecated, or abandoned in the built, or unbuilt, environment.

The third and final part of the book, "Reviving Practices of Repair and Reuse," returns, from different viewpoints and disciplines, to focus on the problem of our gradual and uneven return to repair and reuse, not only as social practices, but also as a form of environmental politics at personal, local, and collective scales. The first, by Tim Cooper and Giuseppe Salvia (Chapter 8), examines the recent history of repair within the larger European context and details the struggles of consumers, traders, and other groups, as well as government agencies, to revive repair as a viable business activity of more visible, and greater, material, environmental, and social value.

The next essay, by Kim Fraser (Chapter 9), is based on a research project intended to see whether it is possible for textile designers to more directly intervene in the environmentally disastrous market for 'fast fashion' that is cheap, new, large-batch clothing. Despite the obvious difficulties involved in remaking clothes from existing ones, the essay explores how this might occur and whether it is possible, or economically feasible, to remake at least some kinds of clothing from the cast-offs of others, without 'downcycling' the materials concerned.

The last essay in this third part, and the penultimate in the book, is a discussion of green waste and composting by Vivienne Waller, Linda Blackall, and Peter Newton (Chapter 10). This looks at how the production of compost, an elemental and very ancient practice of reuse, can be revived, and reinserted into various modern urban settings. As they explain, the environmental and social gains apparent from these types of interventions are potentially substantial, a prime example of what is now termed the Circular Economy (Webster 2015; Crocker et al. 2018). Again, this essay makes the point that the changes required are often small but can have substantial impacts on the levels of waste generated, and on the people involved in generating them.

The chapters in this volume detail just a few examples of the practices of reuse that can now be observed across the many and varied areas of creative endeavor, in this present stage of consumer capitalism. While the extent to which they might succeed in 'subverting consumerism' remains to be seen, they clearly represent a broader questioning of the values and practices that have enabled consumerism to flourish, and demonstrate the potential for, and emergence of, alternative pathways forward.

References

Alexander, C. and Reno, J. (eds) (2012) *Economies of Recycling: The Global Transformation of Materials, Values and Social Relations*, London: Zed Books.

Appelgren, S. and Bohlin, A. (2015) "Growing in Motion: The Circulation of Used Things on Second-hand Markets", *Culture Unbound: Journal of Current Cultural Research*, vol 7, no 1, pp. 143–168.

Aytac, D. O., Arslan, T. V. and Durak, S. (2016) "Adaptive Reuse as a Strategy Toward Urban Resilience", *European Journal of Sustainable Development*, vol 5, no 4, pp. 523–532.

Baum, M. and Christiaanse, K. (eds) (2012) *City as Loft: Adaptive Reuse as a Resource for Sustainable Development*, Zurich, Switzerland: GTA Verlag.

Bourriaud, N. (2002) *Relational Aesthetics*, Dijon, France: Les Presses du Reel.

Brace-Govan, J. and Binay, I. (2010) "Consumption of Disposed Goods for Moral Identities: A Nexus of Organization, Place, Things and Consumers", *Journal of Consumer Behaviour*, vol 9, no 1, pp. 69.

Cherrier, H. (2010) "Custodian Behavior: A Material Expression of Anti-Consumerism", *Consumption, Markets and Culture*, vol 13, no 3, pp. 259–272.

Cohen, L. (2004) *A Consumers' Republic: The Politics of Mass Consumption in Postwar America*, New York: Vintage Books

Cohen, M. J. (2011) "(Un)sustainable Consumption and the New Political Economy of Growth", in K. M. Ekstrom and K. Glans (eds) *Beyond the Consumption Bubble*, London: Routledge.

Cooper, D. and Gutowski, T. (2015) "The Environmental Impacts of Reuse: A Review", *Journal of Industrial Ecology*, pp. 1–19.

Crocker, R., C.P. Saint, G. Chen and Y. Tong (eds) (2018) *Unmaking Waste: Towards the Circular Economy*, London: EmeraldInsight.

Dauvergne, P. (2008) *The Shadows of Consumption: Consequences for the Global Environment*, Cambridge, MA: MIT Press.

Dauvergne, P. (2010) "The Problem of Consumption", *Global Environmental Politics*, vol 10, no 2, pp. 1–10.

Dawdy S.L. (2010) "Clockpunk Anthropology and the Ruins of Modernity", *Current Anthropology*, vol 51, no 6, pp. 761–793.

Dinnin, A. (2009) "The Appeal of Our New Stuff: How Newness Creates Value", *Advances in Consumer Research*, vol 36, pp. 261–265.

Dittmar, H. (2007) "The Costs of Consumer Culture and the 'Cage Within': The Impact of the Material 'Good Life' and 'Body Perfect' Ideals on Individuals' Identity and Well-Being", *Psychological Inquiry*, vol 18, no. 1, pp. 23–31.

Duflou, J. R., Van, O. J. and Dewulf, W. (2013) "Second Thoughts on Preferred End-of-Life Treatment Strategies for Consumer Products", in I. S. Jawahir, S. K. Sikdar and Y. Huang (eds) *Treatise on Sustainability Science and Engineering*, Dordrecht, Netherlands: Springer, pp. 19–29.

Evans, D. (2011) "Thrifty, Green or Frugal: Reflections on Sustainable Consumption in a Changing Economic Climate", *Geoforum*, vol 42, pp. 550–557.

Fry, T. (2009) *Design Futuring: Sustainability, Ethics and New Practice*, Oxford: Berg.

Gibson-Graham, J. D. (2008) "Diverse Economies: Performative Practices for 'other worlds'", *Progress in Human Geography*, vol 32, no 5, pp. 613–632.

Graeber, D. (2012) "Afterword: The Apocalypse of Objects: Degradation, Redemption and Transcendence in the World of Consumer Goods", in C. Alexander and J. Reno (eds) *Economies of Recycling: The Global Transformation of Materials, Values and Social Relations*, London: Zed Books.

Grantham, J. (2011) "Time to Wake Up: Days of Abundant Resources and Falling Prices Are Over Forever", [GMO Capital, online], reposted on *The Oil Drum*, www.theoil drum.com/node/7853, accessed May 1 2015.

Gregson, N. and Crewe, L. (2003) *Second-Hand Cultures*, Oxford: Berg.

Gregson, N., Metcalfe, A. and Crewe, L. (2009) "Practices of Object Maintenance and Repair: How Consumers Attend to Objects Within the Home", *Journal of Consumer Culture*, vol 9, no 2, pp. 248–272.

Gregson, N., et al. (2013) "Moving Up the Waste Hierarchy: Car Boot Sales, Reuse Exchange and the Challenges of Consumer Culture to Waste Prevention", *Resources, Conservation and Recycling*, vol 77, pp. 97–107.

Holroyd, A., et al. (2015) "Design for Domestication: The Decommercialisation of Traditional Crafts", *The Value of Design Research, Paris*, April 2015, http://imagination. lancs.ac.uk/outcomes/Holroyd_Cassidy_T_Evans_M_Gifford_E_Walker_S_2015_ Design_Domestication_Decommercialisation_, accessed 1 November 2015.

Kennicott, P. (2017) "Hamburg's Shimmering Ship of the Skies", *The Guardian Weekly*, vol 196, no 26 (2nd June 2017), p. 40.

Lebow, V. (1955) "Price Competition in 1955", *Journal of Retailing*, vol 31, no 1, pp. 5–10.

Liboiron, M. (2013) "Modern Waste as Strategy", *Lo Squaderno: Explorations in Space and Society*, vol 29, nos 9–12, www.losquaderno.professionaldreamers.net/?cat=162, accessed 1 May 2015.

Lipovetsky, G. (2011) "The Hyperconsumption Society", in K.M. Ekström and B. Glans (eds) *Beyond the Consumption Bubble*, London: Routledge.

Lipsitz, G. (1998) "Consumer Spending as a State Project: Yesterday's Solutions and Today's Problems", in S. Strasser, C. McGovern and M. Judt (eds) *Getting and Spending: European and American Consumer Societies in the Twentieth Century*, Cambridge: Cambridge University Press.

Luckman, S. (2013) "The Aura of the Analogue in a Digital Age: Women's Crafts, Creative Markets and Home-Based Labour After Etsy", *Cultural Studies Review*, vol 19, no 1, pp. 249–270.

Luckman, S. (2015) *Craft and the Creative Economy*. London: Palgrave Macmillan.

McNeill, J. R. and Engelke, P. (2014) *The Great Acceleration: An Environmental History of the Anthropocene Since 1945*, Cambridge, MA: Harvard University Press.

Miller, D. (1998) *A Theory of Shopping*, Cambridge: Polity Press.

Pfister, C. (2010) "The '1950s Syndrome' and the Transition from a Slow Growing to a Rapid Loss of Global Sustainability", in F. Uekoetter (ed) *Turning Points of Environmental History*, Pittsburgh, PA: University of Pittsburgh Press.

Ritzer, G., et al. (2012) "Tracking Prosumption Work on eBay", *American Behavioral Scientist*, vol 56, no 4, pp. 439–458.

Schor, J. S. (2005) "Prices and Quantities: Unsustainable Consumption and the Global Economy", *Ecological Economics*, vol 55, no 3, pp. 309–320.

Slade, G. (2006) *Made to Break: Technology and Obsolescence in America*, Cambridge, MA: Harvard University Press.

Smart, B. (2010) *Consumer Society: Critical Issues and Environmental Consequences*, London: Sage.

Soper, K. (2007) "Re-thinking the 'Good Life': The Citizenship Dimension of Consumer Disaffection with Consumerism", *Journal of Consumer Culture*, vol 7, no 2, pp. 205–229.

Strasser, S. (1999) *Waste and Want: A Social History of Trash*, New York: Henry Holt.

Strasser, S. (2003) "The Alien Past: Consumer Culture in Historical Perspective", *Journal of Consumer Policy*, vol 26, no 4, pp. 375–393.

Tatzel, M. (2014) "Introduction", in M. Tatzel (ed), *Consumption and Well-Being in the Material World*, London: Springer, pp. 1–10.

UNEP and ISWA (2015) *The Global Waste Management Outlook*, report edited by J. C. Wilson et al., http://unep.org/ietc/Portals/136/Publications/Waste%20Management/ GWMO%20report/GWMO_report.pdf, accessed 1 December 2015.

Urry, J. (2010) "Consuming the Planet to Excess", *Theory, Culture and Society*, vol 27, pp. 197–212.

Velis, C. (2014) *The Circular Economy: Closing the Loops* (Report 3, ISWA, online), www.iswa.org/fileadmin/galleries/Task_Forces/ISWA_R3_-_Closing_the_loops.com pressed.pdf, accessed 1 September 2015.

Walker, S. (2011) *The Spirit of Design: Objects, Environment and Meaning*, London: Earthscan.

Ward, S. 2012) "Breathing New Life in the Corpse: Upcycling Through Adaptive Reuse", in S. Lehmann and R. Crocker (eds.), *Designing for Zero Waste: Consumption, Technologies and the Built Environment*, London: Earthscan, pp. 247–266.

Watson, M. and Shove, E. (2008) "Product, Competence, Project and Practice: DIY and the Dynamics of Craft Consumption", *Journal of Consumer Culture*, vol 8, no 1, pp. 69–89.

Webster, K. (2015) *The Circular Economy: A Wealth of Flows*, www.ellenmacarthurfoun dation.org/publications/the-circular-economy-a-wealth-of-flows, accessed 1 July 2015.

Wieser, H. (2016) "Beyond Planned Obsolescence: Product Lifespans and the Challenges to a Circular Economy", *Gaia*, vol 25, no 3, pp. 156–160.

Young, R. A. (2012) *Stewardship of the Built Environment: Sustainability, Preservation and Reuse*, Washington, DC: Island Press.

1 Acceleration, consumerism and reuse

A changing paradigm

Robert Crocker

Introduction

Reuse is all around us. It is now so universal that it has become increasingly privileged in art, fashion, design, and architecture, becoming so widely used as to become almost invisible. No one particularly notices, for example, that in the local café the 30-year-old proprietor has found an old wooden beam and turned it into a defining feature of the bar where the coffee is served, its weather-worn surface offset by some handmade Spanish tiles he has bought online (Manzo 2014). In another example, in a café not far from where I live, old coffee making implements hang from the wall along with an old racing bicycle, suggesting the owner's proud Italian origins and former devotion to the sport of competitive cycling (CBTB).

Suggesting a sense of continuity and place, of cultural difference and "situated identity" (Brinkman 2010), older objects such as these are placed deliberately to be seen and noticed. They can play many roles, including symbolic reminders of place, history, and cultural identity (Sandino 2004; Miller 2008). They can also tell some personal story, some point of difference in origin, or express a commitment to environmental values. However, in the same cafés that one might find such displays of old things, take-away coffee will still be served, typically in unrecyclable plastic-lined paper cups with plastic lids, suggesting a seemingly contradictory commitment to speed, to accelerated modernity, to getting things done more quickly and efficiently, even if they are pleasurable, and probably deserve more time than take-away allows for (Rosa 2003; 2010). So, in the café with the old coffee making equipment and bicycle, I regularly sit and drink my coffee whilst watching the police, ambulance, and fire service people show up in the morning on their way to work, having pre-ordered their take-away coffees online, using an app the patron has especially installed next to the till. Indeed, in this thriving business, two large new Italian machines are going all morning, with one entirely devoted to this take-away trade (CBTB; Rosa 2010; Rosa 2013).

In this chapter I want to look more closely at this now common contradiction, between fast and slow, between presenting – and representing – the past through reuse in a vast range of objects and images, which nevertheless sits beside the ongoing rush of consumerism, and a quest for the 'latest and best', and for doing things faster and faster (Rosa 2010). I will argue that the recovery and reuse of so many older things, from whole buildings to interiors, from the restoration of old bicycles and scooters to the sale of 'vintage' clothing and furniture, has become an important component of a popular, designed response to our environmental problems, and particularly a counter to the technological and social acceleration now evident in today's consumerism (Appelgren and Bohlin 2015). This preference for reuse is often inchoate, vaguely aesthetic and emotive, typically lacking any technical understanding of how engaging in specific acts of reuse might 'reduce carbon' or save particular virgin materials or measurable energy. However, such activities are often self-initiated and thus can also act as a form of 'prosumption', combining production – and design – with consumption and use in a way that creates something unique and personal in an otherwise often bland, mass-produced built and manufactured environment of globally-made substitutable parts (Gregson and Crewe 2003; Ritzer and Jurgenson 2010; Gregson et al. 2013).

While this revival of reuse should be rightly welcomed by environmentalists for its impact on reducing the resources and energy required to make new things from new materials (Castellani Sala and Mirabella 2015; Cooper and Gutowski 2017), I will argue it can also significantly influence how the surrounding world of the new and old, of the natural and human-made, is understood, engaged with, interpreted, and valued (Guiot and Roux 2010; Gregson et al. 2013; Appelgren and Bohlin 2015). Reuse is thus of significance in and of itself as a cultural phenomenon of our times, and not only as a technical solution to an evident but entrenched environmental problem. As I hope to show, it is in part a creative and adaptive reaction to acceleration itself, to the increased speed and mobility, and accompanying 'hyper-consumption', that now characterises so much of modern life (Lipovetsky 2011).

Consumerism and escalation in the throwaway society

Since the late 1960s 'consumerism' has been repeatedly referred to and defined, with some slight variations, as a "way of life and state of mind" in which various acts of consumption themselves seem to become the individual's "way to self-development, self-realization and self-fulfilment" (Benton 1987, p. 245, in Goodwin 1997, p. 3; and see Dittmar 2007; Tatzel 2014; Smart 2010, pp. 8 ff.). While the pursuit of individual fulfilment and self-expression through consumption has a long history, it is only in late eighteenth- and early nineteenth-century industrial America and Europe that active participation in such activities in a consumer society became possible for more than the wealthy (Kroen 2004; Berg 2005). More widely adopted in the early twentieth century, mass-production for mass-consumption first took on its more expansionary and escalatory form from the 1950s as modern democracy and mass-consumption became more closely entwined (Cohen 2004; Meikle 2005; Smart 2010). In this period, there was a unique coincidence of factors that favoured mass-consumption's expansion in the West, and thus the 'downward' spread of consumerism, including the availability of abundant resources and oil for fuels, chemicals and plastics, and the underlying imperative to convert wartime economies to peacetime goals. This engaged more citizens in economic activities that seemed to support democracy, a combination of factors the environmental historian, Christian Pfister (2010) terms "the 1950s Syndrome" (see also McNeill 2010; McNeill and Engelke 2014). In this transformation, the professionalization of waste collection and recycling also played its part in encouraging a more rapid circulation of goods, from production to use and discard (Cooper 2009; Dauvergne 2010; Liboiron 2013a).

However, after the mid-1970s, the collapse of the post-war economic consensus and a new phase of globalization increased the flow and number of more affordable Asian-made goods into Western markets. Increasing computerization also transformed their production, distribution and promotion, increasing volumes and lowering prices of many goods for consumption, at least relative to many incomes (Schor 2005). Steffen, Crutzen and McNeill (2007), in their essay on the *Anthropocene*, refer to this increasing intensity and spread of consumption, at first in the West and then throughout the developing world, as "the Great Acceleration," for it was in this period that increasing rates and volumes of consumption can be seen to have become an escalatory global dynamic, especially from about 1980, and one that necessarily resulted in a dramatic rise in greenhouse gas emissions (Pfister 2010; McNeill and Engelke 2014).

More volumes of goods, including more fast-moving and single-use products, were now being made more cheaply for more people than ever before, who were also being encouraged to buy more, or upgrade what they had bought not so long ago, more frequently. This increasing technological efficiency, greater mobility, and adaptability has been closely tied to an overall increase in consumption speeds and volumes: as prices fall, more people purchase what they can afford, both to 'keep up' with their neighbours and to experience what they previously could not afford (Campbell 2015). Mobile phones can be used to illustrate this point: on the first day of sales in 2010, of the 1.5 million Apple iPhones sold, around three quarters of these went to people already in possession of an iPhone, presumably a working one (Kim and Paulos 2011). Retention rates across many categories of products have steadily fallen during this era of the "great acceleration" (Van Nes 2010; Evans and Cooper 2010). Furniture, kitchens, shop fittings, office and home interiors, and whole buildings, are now being replaced or upgraded more frequently than they were thirty years ago, and many seemingly sooner than even their makers are able to predict (Bakker et al. 2014; Campbell 2015; Wieser 2016).

In many domains, greater durability and longer retention rates would clearly benefit the environment (Evans and Cooper 2010; Bakker et al. 2014). However, consumer preference is in part driven by increasing affordability and universal availability, and also by the promotion of time- or energy-saving products such as electric leaf-blowers, which end up replacing the much more sustainable, but seemingly slower, broom. To this problem can be added that of a growing list of short-lived, often disposable, single-use plastic and card products and packaging, goods that are either unrecyclable or technically recyclable, but in practice only occasionally recycled (Dauvergne 2008; MacBride 2013; ISWA 2015; EMF 2016). These products are high volume and low cost, and so soon become part of an unmanageable global waste stream, contributing directly to environmental degradation and global emissions (Meikle 1997; ISWA 2015).

The lifespans of potentially durable goods can be shortened through various kinds of obsolescence (Slade 2006; Guiltnan 2009). 'Technical obsolescence' is familiar to most, where a part is designed to fail, for example in a toaster or kettle, with the user then pushed to buy another rather than spend money on repair (see Cooper and Salvia, this volume). Cheaper substitute materials, such as chromed plastics used in many household appliances, and laminated MDF and chipboards in furniture, are well-known examples of another version of this type of obsolescence. While the use of such materials might lower prices, they typically pose environmental risks at the end of their life, a cost transferred to the environment, and, eventually, to others (Bartels et al. 2012; EMF 2016). More durable, longer-lived timber furniture, for example, can greatly benefit the environment since its environmental load occurs almost entirely at the beginning of its life and not in its use phase (Cooper and Gutowski 2015). Kitchens, if made well, could last decades, but are now replaced within eight years, and sometimes much sooner (Parrott et al. 2008). Similarly, office buildings, which could last over 100 years, are now typically demolished and replaced in half this time or less (Skelton and Allwood 2013).

Thus, not only has the total volume of goods entering people's lives tended to increase in response to specialized, supposedly time-saving needs, but more frequent upgrading has generally multiplied the impacts of this larger volume and variety of goods (Bauer et al. 2012; Campbell 2015). Barriers to retaining possessions for longer periods, to 'not upgrading', are also much stronger than they have ever been, with these often falling into the category of marketed or visual obsolescence. For example, at the end of their contract mobile phones typically lack insurance coverage, thus encouraging consumers into another contract, enticed with a 'free' phone to upgrade (Guiltnan 2009; Crocker 2012; Wieser 2016). Since repair can involve additional delays and added expense, this type of rapid replacement soon becomes the easiest and safest option, especially for electronic goods (see Cooper and Salvia, this volume).

In this accelerative expansion of consumption, a more rapid disposal of waste becomes an opportunity to sell more products (Dinnin 2009; Liboiron 2013a; Campbell 2015). Focused on the transaction, the fate of the prematurely wasted products becomes a matter for government, or distant others, to deal with and not the manufacturer or the consumer. Literally, this prematurely wasted material becomes 'somebody else's problem' (Crocker 2016).

Deception, lock-in, and post-caution

Since most producers or manufacturers are now under such pressure from the market to produce more goods more quickly, creating waste and pollution are typically treated as external to the requirements of this primary task (Princen 2002; Clapp 2002; Dauvergne 2010). Most plastics, for example, once disposed of, have been known for some time to leak their additives into the environment, becoming persistent pollutants on the land, and in the oceans, but this has had very little impact on plastics manufacturing (Meikle 1997; Eriksen 2014; Glaser 2015). Despite widespread industry and environmental concern, knowledge of plastic's serious environmental impacts has had only a marginal effect on the production and marketing of plastics, which continues to grow in volume and complexity (Liboiron 2013b; Eriksen 2014).

To encourage consumers to accept such routinized externalisation of environmental costs, media-based deception plays an important role in overwriting or distorting the environmental information available to consumers. In promoting cars, for example, products collectively responsible for up to around 20 per cent of greenhouse gases and vast quantities of toxic waste and pollution, a SUV can now be advertised as "nature's friend" (Rollins 2006), whilst a hybrid car, because of its lower emissions in use (but not in manufacture or disposal), can be presented in striking visual terms as "invisible" to the environment, suggesting it has "zero" emissions (Li 2013). Such gross misrepresentations mask the significantly negative environmental impacts of these products, and the systems enabling and supporting them (Crocker 2013). Rather like the early 'health based' advertising of cigarettes in the 1920s, deceptive environmental advertising is often misconstrued as an exception, typically in the emotive term 'greenwashing', rather than the rule, a rule that is reinforced by its apparent normality. Thus, we no longer expect to be told the truth about what our products might do to the 'distanced' oceans (Liboiron 2013b; Glaser 2015).

Such routinized deception is aided by consumer dependence on many unsustainable products and systems (Sanne 2002; 2005; Dauvergne 2008). For example, once a city's roads become essential to transport services, they cannot be replaced, since so many now depend on them, and the car soon becomes seen as an essential and expected service available to everyone (in theory), its environmental costs deferred or hidden from the system's users themselves (Soron 2009). Dependence on such systems, and the services they provide, generates a 'lock in', or a type of 'sunk-cost effect', that is, a commitment created by the irrecoverable investments made in the past to the creation and maintenance of the system concerned, sometimes over many years (Janssen Kohler and Scheffer 2003; Kelly 2004). Sunk-cost effects involve an overestimation of the benefits of the system concerned, and a corresponding underestimation of its negative impacts (Cunha and Caldieraro 2009), along with a tendency to ignore or deny the value of possible alternatives (Crocker 2016, pp. 77–95).

This combination of acceleration in the cycle of production, use and discard, widespread acceptance and use of deception, and the sunk-cost effects of the systems in use, embed what might be best termed a 'post-cautionary' pathway, from design and manufacturing to disposal, in which most products' negative social and environmental impacts are externalized, concealed, or denied, until such a time when there is no choice but to acknowledge and face up to them (see Princen 2002). I have borrowed this term "post-cautionary" here from John Paull (2007), an Australian forest ecologist who describes a "post-cautionary principle" at work in much environmental decision-making today. Referring to a number of cases, including the extraordinarily unremarked death of 'El Grande', possibly the oldest tree in Australia, through a seemingly routine, error-prone, forest clearing activity a few years earlier, Paull (2007, p. 3) details the kind of instrumental thinking that limits the value of these living 'resources' to their value as timber in consumption. For Paull, this approach is the perverse opposite of the ancient precautionary principle, 'first do no harm', and indeed embodies its opposite, which he defines in these terms:

> Where there are threats of serious or irreversible damage, the lack of full scientific certainty shall be used as a reason for not implementing cost-effective measures until *after* the environmental degradation has actually occurred [my italics].

Acceleration and deceleration

Post-cautionary approaches to design and production are now reinforced by the time-pressures involved in the increased speeds of the cycle of production and consumption, for this 'acceleration' tends to encourage all kinds of process or technical short-cuts (often rationalised as 'efficiencies') to get the product to market more quickly (Lipovetsky 2011). This increase in pace is not only seen in technological and process changes and efficiencies, but also in their accompanying psychological and social effects (Tomlinson 2007). The dominance of the plastic-lined paper throwaway coffee cup, for example, is suggestive of how this works: the thought that one might save a few minutes by taking away a coffee, is sufficient to drive more and more to take up this environmentally damaging practice; Starbucks is said to need over eight million cups every day in the US alone, and their business model now depends on providing seating for only a small percentage of their 'walk-through' custom (Quin 2016).

The social theorist Hartmut Rosa (2003; 2010; 2013) provides a useful frame through which this increasing acceleration can be viewed, emphasising an interdependence between technological, psychological and social forms of acceleration, and their reinforcing feedback effects. Technological acceleration for Rosa boosts social acceleration, which results in a transformation of spaces and relationships (2003, pp. 6–7; 2013). For example, the car has increased personal mobility and choice, and this change has resulted in a series of new social practices, which in their turn reinforce or embed a greater car-dependence, or lock in (Sanne 2002; Sanne 2005; Soron 2009; Crocker 2016, pp. 77–95). The economic dimension of such powerful technological and social accelerators is driven and expressed through consumption, not only of cars, fuels, roads, etc., but also of many related services, such as shopping centers and suburban residential and commercial developments, what might be termed 'carscapes' (Morrison and Minnis 2013).

While technological and social acceleration reinforce each other, developing side by side, this has a significant individual psychological dimension in an increase in the pace of life itself (Rosa 2013, pp. 71 ff.). Rosa argues after the philosopher, Hermann Lübbe (2008), that social acceleration gives rise to a "contraction of the present" as the realms of experience (the past) and expectation (the future) are brought closer together in our consciousness (Rosa 2003, p. 7). This is because the present is now filled with more choices and things to do (Garhammer 2002; Glorieux et al. 2010), making it harder to remember and draw upon past experience, in order to respond to what might be expected in the immediate future (Rosa 2003, p. 7). Thus, while apparently saving time, faster technologies of mobility and communication increase choice, but also tempt people to try and do more, and more quickly, within the time available (Rosa 2003, pp. 9–10; 2010).

Drawing on a number of social theorists including Marx, Simmel, Weber and Luhrmann, Rosa (2003; 2013) presents acceleration as being driven primarily by three interacting "motors": an economic motor which drives technological change, expressed in the idea, for instance, that "time is money" (2003, p. 12), a structural motor that results in a continuous expansion of complexity or "functional differentiation", expressed through social acceleration where more increasing demands drive a recourse to more short-cuts, many of them embodied in new devices or routines (2003, pp. 14–15), and a psycho-social or "cultural motor", which he terms a "eudaemonic" impulse (a desire for the 'good life'), which results in an acceleration of the experience of the pace of life itself, and often a longing to escape the pressure this brings in its train (Rosa 2003, p. 13). The individual's desire for the good life, or a seemingly better one, is made possible through the many choices that acceleration itself makes possible (Rosa 2013, pp. 174 ff.; and see Dittmar 2007). However, as more choices appear, there is less time to enjoy them, so more time-saving devices or routines must be adopted to get 'there' faster and more efficiently (Rosa 2003, pp. 13–14).

One of the more interesting features of Rosa's argument is thus his emphasis that acceleration typically generates an opposing tendency towards deceleration, to escaping the pressures that acceleration seems to apply to the individual (2003, p. 5 and pp. 14 ff.). In his essay, Rosa distinguishes between five forms of deceleration (pp. 15–17), including natural limits such as the biological limits of our bodies and minds; surviving oases of deceleration, such as the slower pace and much longer times required to make whiskey; an enforced deceleration, which might be the unintended consequence of acceleration itself such as is commonly experienced in traffic jams; various forms of intentional opposition to acceleration, where an older, slower practice might be deliberately retained or revived, such as playing a musical instrument, or listening to one being played; and finally, a collapse or stasis created by too much choice and acceleration, a postmodern end to acceleration itself. It is Rosa's fourth category, intentional opposition to acceleration in and through various voluntary forms of deceleration, that I am mainly concerned with here.

Desynchronization, nostalgia and memory

Deceleration cannot be fully understood simply in terms of a reactive outcome of acceleration, of rapid technological innovation or economic and social change, but as a moment within acceleration itself, a form of "desynchronization" (Rosa

2003). Indeed, this tendency towards desynchronization becomes especially visible during periods of intensive technological change or acceleration, and can even shape how these accelerative forces are understood and perceived (Viera 2011). For example, at the height of industrialisation in the Victorian era and the introduction of railways, steamships and factory production, a deep fascination with the medieval past and its seemingly greater certainties, its craftsmen and peasants, gripped the English imagination (Evans 1988; Elliott 2000; Alexander 2007). This romantic return to the past often masked and accompanied a critical engagement with the present, with a desire to reform or change the worst of industrialism and the factory system by wistfully comparing it to what seemed to have gone before (Breton 2002; Kinna 2006). William Morris's supposed "revival" of crafts, many of which were still very much alive in his day (Adamson 2013, pp. 181–184), is a good example of how cultural renewal and innovation can be framed referencing memory in nostalgic, ideological terms, and making use of a recalled past (however historically inaccurate) to counter the accelerative and destructive present (Dawdy 2010). Morris's refusal to go into the Great Exhibition in 1851 because of its alleged ugliness, at the age of only seventeen, has been read as the first statement of his lifelong ideological commitment to both a rejection of the factory system, and an assertion of the independence of design in craft-based production (Adamson 2013, pp. 181–184).

Such imaginative returns to the past are often referred to negatively as expressions of 'nostalgia', a word which has retained its medical, supposedly pathological, origin (Boym 2007, p. 7). In her engaging study of the subject, however, Svetlana Boym (2001; 2007) usefully distinguishes between two tendencies within nostalgia, a conservative "restorative nostalgia," and a more creative "reflective" one (2007, p. 8; Boym 2001, pp. 41–55). While the first suggests a desire for a complete restoration of what was allegedly some significant feature of the past, and emphasises 'nostos' (Greek for 'home') over longing itself, the latter, "reflective" nostalgia, emphasises longing or 'algia', from the Greek for the bittersweet longing that memory can evoke (Boym 2007, p. 8). The most extreme expression of the first occurs in reactive, conservative, religious and political movements, and institutions, or in individuals who believe in attempting to 'restore' a place and time from the past itself. Referring to the grandiose architecture of seventeenth century Catholicism, of European Fascism and Soviet Stalinism, Boym (2001, pp. 41–48) emphasises the potentially painful and violent consequences of restorative visions of a 'remembered' past. What is longed for in the past in restorative nostalgia can become an absolute truth, an ideal that 'should' be restored – often in militant opposition to some designated enemies – a "restoration" that is assumed, falsely, to cure the ills of modernity (Boym 2007, p. 13).

For Boym, "reflective nostalgia," on the other hand, is more creative and adaptive, and often ironic. It has no need to try and return to the past itself, since its proponents acknowledge that this "home" in time is no longer accessible, except in reflective memory (Boym 2007, p. 15 ff.; and see Ritivoi 2002). This creative and potentially transformative form of nostalgia is epitomised for Boym by the detailed, memory-rich creations of great modern writers like Proust and Nabokov, and in much great modern art, design, and literature. As Boym emphasises, reflective nostalgia savours "details and memorial signs, perpetually deferring homecoming itself" (2001, p. 49). Indeed, it is the gap between what is remembered and the present that stimulates the reflective work of art, literature, or design. For it is the "defamiliarization and sense of distance that drives [the reflective nostalgic] to tell their story, to narrate the relationship between past, present and future" (2001, p. 50). Boym's argument again highlights that memory, in the face of the rapid changes generating modernity's ongoing 'forgetting' (Connerton 2009), remains a crucial creative resource to enable a fuller acceptance, and integration, of rapid change towards a sense of renovation and renewal.

In design, too, there are tendencies within this return to the past, from direct imitation to the creative framing of a story, image, experience, or object, within a theatre provided by a shared memory from the past. To take two examples, roughly coincident with Boym's two tendencies, a restorative and imitative tendency in design can result in the complete restoration (or rebuilding) of a structure, and its elevation into a monument, perhaps representing some important distinguishing character or origin which a community wishes to celebrate (Brett 1996; Connerton 2009). It can also result in the masking of otherwise modern objects or buildings, to make them look old, as in the mansions built by the industrial moguls of the nineteenth century, many of which were designed in the style of more famous older structures, from Renaissance palaces to eighteenth-century French chateaux, with many even incorporating antique ceilings, panels, windows and doors, but also, of course, the latest in modern comforts like central heating, hot water plumbing, and electric lights (Waddesdon 2016; and see Elliott 2000). This restorative tendency might indeed, as Boym argues, be associated with power and authority, as can be seen, for instance, in the grand neoclassical entrance to many nineteenth century banks, and in many of the grandiose London houses the wealthy built over the last two centuries (Strong 1996; Cornforth 2000; Crocker 2015).

However, as in Boym's distinction, there are many instances of a more creative reflective use of memory, evident in many modern works of art, architecture and design, and also in various narrative expressions in popular culture, including novels, films and computer games (Pugh and Weisl 2013). J.R.R. Tolkien's *Lord of the Rings* (Gaunt 2003) is a case in point. At one time, it was one of the most popular books of the sixties generation, despite starting as a 1920s, post-World War 1 tale, based on some poems the author had written under the influence of William Morris (Gaunt 2003). While its carefully crafted 'authenticity' might fool us into imagining that this is a work of restorative nostalgia in its intention, its distance in time, and implicit acknowledgement of this distance, emphasises its spiritual descent from Morris, another great reflective nostalgic (Breton 2002; Delveaux 2006). The remembered past in works such as this provides a mental space that opens up the possibility of other sensual worlds, a richness and imme- diacy of experience suggestive of deceleration's continuing seductive fascination.

Reflective nostalgia and the consumption of experience

In a more recent postmodern example, that also draws directly upon Boym's concept of reflective nostalgia, Bjorn Schiermer (2014) writes of the hipster phenomenon, and the hipsters' nostalgic relationship to 1950s and 1960s kitsch objects and fashion, as "a shared investigation into the possibilities, potentialities and sensibilities of past aesthetical universes" (p. 176). Drawing on Boym's ideas about nostalgia, Schiermer emphasises the hipsters' self-conscious, and often ironic "redemption of the past," elevating their discoveries into icons of a new hybrid style (Schiermer 2014, pp. 169–172). The case of the hipsters also brings into focus the changes that have taken place since the early 1980s, and how these have changed not only how the remembered past is understood outside the constraining but more ordered narratives of formal history (Rosenzweig and Thelen 1998), but how the self comes to inhabit its imagined memories, in an accelerated post-industrial world of continuous "situational" change, where the past is understood, and learnt about, through popular visual media and film rather than through formal study (Rosenzweig and Thelen 1998; Leone 2015).

Citing Rosa (2003), and also the ethnographic studies of Richard Sennett (1998; 2007) on the dramatic changes in the culture of work in post-industrial America, the psychologist Svend Brinkman (2010) describes the social dimensions of three 'ages' of the self over the last two hundred years, from the nineteenth-century bourgeois notion of 'character', to the twentieth-century's post-Freudian 'personality', to the post-modern and more mobile 'situated self' (see also Dawdy 2010). Within a now dissipating field of relatively fixed and hierarchic social and economic relationships, the individual's personality, which once seemed a cogent way of explaining difference, and the vicissitudes of social interaction, has now been eclipsed by a 'situational identity', reflecting the ever-greater mobility, responsiveness and flexibility required of individuals today. Like Schiermer's hipsters, these must adapt to a life of continuous social and material change, within a shifting network of relationships, diminishing economic opportunities, and more unstable living arrangements (Brinkman 2010). In this new, more intensively accelerated world, narratives of identity are to be discovered through a collage of personal mnemonic associations, accumulated through the media's talismanic objects or images, and placed together or curated – in the real world or online – for stabilizing personal and social effect (Leone 2015).

In terms of consumption, this shift from personality-focused discourses to identity-focused ones, is expressed through a change in consumption priorities, from attaining possessions that may augment an individual's sense of identity and social position over longer periods of time (Miller 2008), to attaining more immediate access to the consumption of a valued experience itself (Manzini 2002; Brinkman 2010).

In this case, a fascination with 'situational identity' can be seen to begin earlier than the hipsters, in the more experimental identity politics, environmentalism, and self-exploratory ethos of the counterculture of the 1960s (Binkley 2003; Rome 2003; Turner 2005; 2006; Kirk 2007; Suri 2010). Opposing consumerism, and protesting the environmental destruction that it contributed to, the counterculture was linked to a nostalgic 'back to nature' communitarianism, and a do-it-yourself ethos of making, building, and growing your own, typically to a mythical historical model, a kind of prosumption or reversion to a DIY and DIT ('do it together') production 'for' consumption, that is still in evidence in many pro-environmental initiatives today (Rome 2003; Turner 2006; Conn 2010; and Chiveralls, this volume). The thread of self-liberation, environmentalism, and communalism, for example, runs through Stewart Brand's *Whole Earth Catalog: Access to Tools* (Brand 1968; Kirk 2007), where one could learn how to build a mud-brick home or even a geodesic dome, grow food, spin yarn and make textiles for clothing, make and use compost, or generate power from the wind, or create other forms of 'appropriate technology' (Binkley 2004; Turner 2006; Kirk 2007; Conn 2010). The *Whole Earth Catalog* not only included instructions and descriptions of how useful things might be made or used, and even where one might buy the parts for making these, but also philosophical justifications for why 'doing your own thing' – that is making or producing what one might need to consume – again through a return to selected practices of the past, was best for the individual, the environment, and the surrounding society (Turner 2006; Kirk 2007; Conn 2010).

From slow food to sustainable design

Such imaginatively 'remembered' and revived practices from the past seem intended to subvert the often-rapid destruction of memory imposed by modernity and its accelerative tendencies (Connerton 2009; Dawdy 2010). In the various movements referred to above, art, design and craft became linked to a remembered past, whose imagined existence was referred to, in order to justify the fashion of making and doing in an (allegedly) past style, rather than simply consuming the mass-manufactured products of industry alone (Adamson 2013; Luckman 2013; 2015). These exemplary movements also share a preference for 'prosumption', that is, making or producing something which can then be used or consumed (Ritzer and Jurgenson 2010). It is the user's engagement in the 'lost' practice that transforms and intensifies its experience and meaning, and makes it seem more real in the present. So, William Morris and his followers repeatedly claimed that the crafts they were "reviving" had died, despite contemporary evidence to the contrary (Adamson 2013, pp. 181–185).

In this process of ongoing, and often experimental, recovery or memory work, the past practice, and sometimes associated objects or materials, are rediscovered for reuse, and then folded into the present, both as a means of engaging more fully in the sensual experience of making-doing to consume, and as a means for justifying why this alleged practice, excavated from the past, is to be valued so highly. In this way, a situated identity becomes grounded, and made seemingly more real, against the tide of a landscape of continuing technological and social change. Whether what is rediscovered, and reused, is playing an old and neglected instrument, throwing a ceramic pot, growing a garden of herbs and vegetables, or building a tiny house, the imagined past as a reference point is reassuring, since it seems to embody what has not been entirely erased, destroyed, or replaced, in the remorseless destructiveness of the advance of modernity (Ritivoi 2002; Connerton 2009; Dawdy 2010).

As John Tomlinson (2007) suggests, in reference to the Italian Slow Food movement, such rebellions against modernity's social and technological acceleration are creative, adaptive responses "to a complex and value-ambiguous cultural condition," a "complicated lifestyle 'problem' to be confronted, rather than a goal to be achieved" (2007, p. 148). This "lifestyle problem" is typically entangled with some disliked or feared form of acceleration (Rosa 2013, pp. 251–276). The aim of movements such as Slow Food, is thus not so much a conservative 'restorative' opposition to what is new that might be disliked, such as fast food or the cruelties of chemical agriculture, but a more 'reflective' demonstration, of how to live without these fast and seemingly corrupt practices, and to consciously disengage from them, through returning to a more fulfilling relationship between person, practice, and environment. As Carlo Petrini (2001), the founder of Slow Food, has himself emphasised, the aim of Slow Food is not just to right some specific wrong, but to assert "the right to pleasure" itself in the domain of food, in its production, preparation, and consumption, a right that is now ignored or trampled on, in the rush to exploit both producer and consumer through fast food systems and the agribusinesses that support them (Petrini 2001; Manzini and Tassinari 2013).

In their essay on the "sustainable qualities" underpinning design for social innovation and sustainability, Manzini and Tassinari (2013) turn to Slow Food, and its right to pleasure, as a model for sustainable design to follow. They argue, like the proponents of Slow Food, that designers should work to create new forms of communal relationship between people, things, and environments, and, in and through this work, strive to generate new attitudes towards time, to counter a now common emphasis on time's alleged scarcity, and also its supposed equivalence to money. Manzini and Tassinari also draw attention to how Slow Food has called for a revaluation of meaningful, skilled, and engaging work, along with the importance of place, localized collaborative production and consumption, and DIY prosumption. They also emphasise its stimulation of alternative forms of exchange that do not depend solely on monetary transactions (Manzini and Tassinari 2013, pp. 225–229; and see Gibson-Graham 2008; Seyfang 2009).

These principles, which they summarise as "deep relationships, ecology of time, meaningful work, collaboration by choice, human scale, cosmopolitan localism and enriching complexity" (2013, p. 229), typically also involve the folding of a 'remembered' past practice into the accelerated practices of the present. For the aim of sustainable design, like Slow Food, is not simply a response to the abstract need to reduce carbon, energy, or waste, but to rediscover and value a more authentic relationship between people, products, and places, and to redeem these from the rapid destructiveness that an accelerated modernity embodies (Dawdy 2010). This is to be accomplished through a creative engagement with what already exists, and an open rejection of consumerism's obsessive quest for the new, the latest and the best (Dinnin 2009; Walker 2011; Campbell 2015). In this quest, design is to be used to promote a deeper and more authentic experience of pleasure in daily life. This can strengthen or renew a relationship with community and place, and encourage alternative forms of exchange, and more sustainable forms of possession and work (Manzini and Tassinari 2013; and Black and Cherrier 2010).

Reuse and the co-creation of value

Within this approach to sustainability, reuse in more recent design emerges not just as a consciously functional and reductive environmental program (Castellani Sala and Mirabella 2015), but as an aesthetic, ethical, and co-creative, response to a destructive and accelerated consumerism (Walker 2011; Crocker 2016). Consumerism has made the new the measure of all value, and turned the old,

Figure 1.1 Main Assembly Building, with Mitsubishi symbol, Tonsley (2017)

Photo: Robert Crocker.

Figure 1.2 Gardens under Roof, Main Assembly Building, Tonsley (2017)

Photo: Robert Crocker.

prematurely, into waste (Campbell 2015). Instead, working with existing social and material relations, the designer can revalue, reshape, strengthen and extend what exists, accepting this as potentially sufficient, or perhaps nearly so (Baum and Christiaanse 2012; and Lee, this volume). While some designers might seek to explicitly advance an environmental agenda, many now include a social dimension, that seeks to encourage communal, collaborative relationships, and new synergies and possibilities, that the relative isolation and separation of many 'fast' modern design programs have tended to undermine (Ward 2012; Christiaanse 2012; Baum and Christiaanse 2012).

In one remarkable Australian example of this type of regenerative design, a government-funded multi-use eco-precinct was planned to replace an old car factory, situated about 10 kilometres south of Adelaide, in South Australia (Tonsley 2017). Originally built to house Australia's substantial Chrysler production line, the factory hit hard times in the 1980s, before being taken over by Mitsubishi. Increasingly automated, it eventually closed in 2008. The surviving Main Assembly Building included one of the largest industrial rooves in the country, spanning some 11 hectares (over 20 acres), a large saw-tooth in steel, asbestos sheets and glass, that the designers, Woods Bagot and Tridente Architects, elected to retain (WAN 2015).

The architects made this decision, in part to enable the creation of a more temperate micro-climate beneath it, since the vast roof provides such an effective shelter from Adelaide's harsh sun, and in part to create a sense of place and continuity, in an otherwise typical landscape of demolished industrial sheds and nondescript seventies office buildings. They also wanted to exploit the more pragmatic opportunities presented by such a vast roof for solar power generation and rainwater collection. They found support for their plan in the heavy costs associated with dismantling such a large industrial structure (WAN 2015; Tonsley 2017).

Approximately two storeys high, and in several places four storeys high, Tonsley's restored steel and glass structure tells a significant local story, a story of

Figure 1.3 Gantry and Atrium, TAFE College, Tonsley (2017)

Photo: Robert Crocker.

post-war reconstruction, industrialization and modernization, eventual decay and abandonment, one that is closely tied to the history of modern Adelaide, as to many other now post-industrial cities (Tonsley 2017; Spoehr 2017). The designers have inserted beneath, or sometimes through, this roof an array of new commercial and office buildings, educational facilities and retail outlets, along with many landscaped social spaces. The Main Assembly Building's aesthetic impact comes from what is more than simply a thoughtful juxtaposition of old and new, for it contains various attempts to embody local memory in its design, with the old gantry and several signs, including one from Mitsubishi, remaining on the worn concrete floor (see Figure 1.3). There are now gardens and micro-forests, water features, seats and social spaces, dotted between the various commercial and educational buildings that this vast roof houses, including a technical college covered largely in glass, whose workshops can be seen exposed to view, another strong element asserting continuity with the past (Tonsley 2017).

It is apparent that Tonsley attempts to engage its visitors in many of the "sustainable qualities" Manzini and Tassinari (2013) write about: it has been built not to a functional, hierarchic agenda, as a new corporate 'campus' might (see WAN 2015; Tonsley 2017), with clearly defined zones of activities and services, but as a blended and dynamic working community arrayed in arcs around a central hub, the Main Assembly Building, with its design deliberately encouraging everyday social interaction. Its educational facilities, for instance, sit beside entrepreneurial technology start-ups, and share services with them, such as a series of cafés and shops, whilst workers and residents also overlap in using the public spaces under the building's vast roof.

Conclusion

In a world overfilled with things, and one which jumbles together for many, things that are needed and things that are not, the reuse of resources, systems, places and people, makes increasing sense, both environmentally and economically (Castellani, Sala, and Mirabella 2015; Cooper and Gutowski 2015). However, reuse also offers a more elusive cultural and social thread to follow in an era of constant and sometimes unsettling, and even personally costly, change, a metaphorical thread that seems to lead back to ways of understanding, making, using and experiencing the world of goods, services and environments that do not require the destruction of the old, its wastage and the production of the brand new, of the 'latest and the best'.

Developments like Tonsley might be seen as innovative against the global tide of lookalike malls, garden suburbs, and vast new industrial estates. Certainly, against the high-consuming Western-style city, such deliberately low-consuming eco-precincts attempt to encourage alternative social and economic activities, to generate new forms of relationship, collaboration, work, leisure and creative exchange, that do not require so much energy, material throughput, and use of scarce resources (Seyfang 2009; and see Gibson-Graham 2008). These new, often multi-functional and blended places might represent the future, but significantly, a future that does not attempt to devalue and erase the past, but instead embraces its memory, folding this into what might appear to be a longer, less contracted present.

In design terms, Tonsley's hybrid blending of the old and new is part of an increasingly global, late modern trend. In this model the old is no longer seen as either so much redundant waste, to be swept aside and replaced more rapidly to generate more value, or something to be carefully restored and renovated to some imagined historical standard, but rather as an opportunity for creative adaptation and expression, to turn something redundant into something useful for the times (Ward 2012). The old object, material, or practice, might be discovered, renewed and folded into the new, but its recovery becomes a dual sign – on the one hand, embodying a critique of the accelerated modern and its consumerism, and on the other, pointing to a richer, more direct, and more sociable and sensual engagement, a "right to pleasure" that does not require a continuously expanding and accelerating production and consumption of the new (Manzini and Tassinari 2013).

It may be that Tonsley's next phase, its large residential development, which has only just begun, will not be as successful as the initial restoration of the Main Assembly Building, and its first three, exceptional, new educational and government buildings it has been integrated with (Tonsley 2017). It may be that its relative isolation from traditional facilities, such as schools, shops, and parks, will afflict its first years of social development. It is possible that the design of this larger area will not reflect the high standards on show now in the Main Assembly Building (Tonsley 2017). However, it is still worth emphasising that such exemplary large-scale displays of reuse have become increasingly influential in the world of architecture and development, and are now influencing the way such large infill developments are designed, planned, and presented (Ward 2012; Christiaanse 2012; Kennicott 2017).

The material advantages of reuse can be readily recognised, and might contribute directly to a more circular economy, since the old, and seemingly redundant and devalued, can now be revalued, and inserted into the new (see Crocker et al. 2018). Such developments might also reduce waste and create financial value, and even increase the employment opportunities that governments in post-industrial states most desire. However, this reinsertion of the old into the new has considerable cultural and social significance, beyond the more limited and objective terms of its functional benefits (Applegren and Bohlin 2015). Linking memory to place, restoring what is more typically erased or wasted, can create a sense of continuity, belonging and relationship, where before there was little or none.

References

Adamson, G. (2013) *The Invention of Craft*, London: Bloomsbury.

Alexander, M. (2007) *Medievalism: The Middle Ages in Modern England*, New Haven, CT: Yale University Press.

Appelgren, S. and Bohlin, A (2015) "Growing in Motion: The Circulation of Used Things on Second-hand Markets", *Culture Unbound: Journal of Current Cultural Research*, vol 7, no 1, pp. 143–168.

Bakker, C., et al. (2014) "Products That Go Round: Exploring Product Life Extension Through Design", *Journal of Cleaner Production*, vol 69, pp. 10–16.

Bartels, B., et al. (2012) "Introduction to Obsolescence Problems", in B. Bartels et al. (eds) *Strategies to the Prediction, Mitigation and Management of Product Obsolescence*, New York: Wiley.

Bauer, M. A., et al. (2012) "Cuing Consumerism", *Psychological Science*, vol 23, no 5, pp. 517–523.

Baum, M. and Christiaanse, K. (eds) (2012) *City as Loft: Adaptive Reuse as a Resource for Sustainable Development*, Zurich: GTA Verlag.

Benton, R. (1987) "Work and the Joyless Consumer", in A. Firat, N. Dholakia and R. Bagozzi (eds) *Philosophical and Radical Thought in Marketing*, Lexington, IN: Heath.

Berg, M. (2005) *Luxury and Pleasure in Eighteenth Century England*, Oxford: Oxford University Press.

Binkley, S. (2003) "The Seers of Menlo Park: The Discourse of Heroic Consumption in the 'Whole Earth Catalog'", *Journal of Consumer Culture*, vol 3, no 3, pp. 283–313.

Binkley, S. (2004) "Everybody's Life is Like a Spiral: Narrating Post-Fordism in the Life-style Movement of the 1970s", *Cultural Studies: Critical Methodologies*, vol 4, no 1, pp. 71–96.

Black, I. and Cherrier, H. (2010) "Anti-consumption as Part of Living a Sustainable Life-style: Daily Practices, Contextual Motivations and Subjective Values", *Journal of Consumer Behaviour*, vol 9, pp. 437–453.

Boym, S. (2001) *The Future of Nostalgia*, New York: Basic Books.

Boym, S. (2007) "Nostalgia and Its Discontents", *The Hedgehog Review*, Summer, pp. 7–18.

Brand, S. (ed) (1969) *The Whole Earth Catalog: Access to Tools,* San Francisco, CA: Portola Institute.

Breton, R. (2002) "Work Perfect: William Morris and the Gospel of Work", *Utopian Studies*, vol 13, no 1, pp. 43–56.

Brett, D. (1996) *The Construction of Heritage*, Cork: Cork University Press.

Brinkman, S. (2010) "Character, Personality and Identity: On Historical Aspects of Human Identity", *Nordic Sociology*, vol 62, no 1, pp. 65–85.

Campbell, C. (2015) "The Curse of the New: How the Accelerating Pursuit of the New Is Driving Hyper-consumption", in K. M. Ekstrom (ed) *Waste Management and Sustainable Consumption: Reflections on Consumer Waste*, London: Routledge.

Castellani, V., Sala, S. and Mirabella, N. (2015) "Beyond the Throwaway Society: A Life-cycle Assessment of the Environmental Benefit of Reuse", *Integrated Environmental Assessment and Management*, vol 11, no 3, pp. 373–382.

CBTB, Coffee by the Beans, www.coffeebythebeans.com.au, accessed 1 December 2015.

Christiaanse, K. (2012) "Traces of the City as Loft", in M. Baum and K. Christiaanse (eds) *City as Loft: Adaptive Reuse as a Resource for Sustainable Development*, Zurich: GTA Verlag.

Clapp, J. (2002) "The Distancing of Waste: Overconsumption in a Global Economy", in T. Princen, M. Maniates and K. Conca (eds) *Confronting Consumption*, Cambridge, MA: MIT Press.

Cohen, L. (2004) *A Consumers' Republic: The Politics of Mass Consumption in Postwar America*, New York: Vintage.

Conn, S. (2010) "Back to the Garden: Communes, the Environment, and Anti-urban Pastoralism at the End of the Sixties", *Journal of Urban History*, vol 36, pp. 831–848.

Connerton, P. (2009) *How Modernity Forgets*, Cambridge: Cambridge University Press.

Cooper, D. and Gutowski, T. (2015) "The Environmental Impacts of Reuse: A Review", *Journal of Industrial Ecology*, vol 21, no 1 pp. 1–19.

Cooper, T. (2009) "War on Waste?: The Politics of Waste and Recycling in Post-War Britain, 1950–1975", *Capitalism Nature Socialism*, vol 20, no 4, pp. 53–72.

Cornforth, J. (2000) *London Interiors: From the Archives of Country Life*, London: Aurum.

Crocker, R. (2012) "Getting to Zero Waste in the new mobile communications paradigm: a social and cultural perspective", in S. Lehmann and R. Crocker (eds), *Designing for Zero Waste: Consumption, Technologies and the Built Environment*, London: Earthscan: pp. 115–130

Crocker, R. (2013) "Ethicalization and Greenwashing: Business, Sustainability and Design", *Design for Business: AGIDEAS Research*, vol 2, pp. 162–175.

Crocker, R. (2015) "The Haunted Interior: Memory, Nostalgia and Identity in the Interwar Interior", in D. Daou, D. J. Huppatz and D. Q. Phuong (eds) *Unbounded: On the Interior and Interiority,* Sheffield: Cambridge Scholars.

Crocker, R. (2016) *Somebody Else's Problem: Consumerism, Sustainability and Design*, Sheffield, UK: Greenleaf.

Crocker, R., Saint, C.P, Chen, G. and Tong, Y. (eds) (2018) *Unmaking Waste: Towards the Circular Economy*, London: EmeraldInsight.

Cunha, M. and Caldieraro, F. (2009) "Sunk-Cost Effects on Purely Behavorial Investments", *Cognitive Science*, vol 33, pp. 105–113.

Dauvergne, P. (2008) *The Shadows of Consumption: Consequences for the Global Environment*, Cambridge, MA: MIT.

Dauvergne, P. (2010) "The Problem of Consumption", *Global Environmental Politics*, vol 10, no 2, pp. 1–10.

Dawdy, S. L. (2010) "Clockpunk Anthropology and the Ruins of Modernity", *Current Anthropology*, vol 51, no 6, pp. 761–793.

Delveaux, M. (2006) "'O me! O me! H I love the earth': William Morris's News from Nowhere and the Birth of Sustainable Society", *Contemporary Justice Review*, vol 8, no 2, pp. 131–146.

Dinnin, A. (2009) "The Appeal of Our New Stuff: How Newness Creates Value", *Advances in Consumer Research*, vol 36, pp. 261–265.

Dittmar, H. (2007) "The Costs of Consumer Culture and the 'Cage Within': The Impact of the Material 'Good Life' and 'Body Perfect' Ideals on Individuals' Identity and Well-Being", *Psychological Inquiry*, vol 18, no 1, pp. 23–31.

Elliott, B (2000) "Historical Revivalism in the Twentieth Century: A Brief Introduction", *Garden History*, vol 28, no 1, pp. 17–31.

EMF (2016) *The New Plastics Economy*, Ellen Macarthur Foundation report, www.ellen macarthurfoundation.org/programmes/business/new-plastics-economy, accessed 1 December 2016.

Eriksen, M. (2014) "Plastic Pollution: The Plastisphere – The Making of a Plasticized World", *Tulane Environmental Law Journal*, vol 27, pp. 153–393.

Evans, S. and Cooper, T. (2010) "Consumer Influences on Product Lifespans", in T. Cooper (ed), *Longer Lasting Products: Alternatives to the Throwaway Society*, Farnham, UK: Gower.

Evans, T. H. (1988) "Folklore as Utopia: English Medievalists and the Ideology of Revivalism", *Western Folklore*, vol 47, no 4, pp. 245–268.

Garhammer, M. (2002) "Pace of Life and Enjoyment of Life", *Journal of Happiness Studies*, vol 3, no 3, pp. 217–256.

Gaunt, J. (2003) *Tolkien and the Great War: The Threshold of Middle Earth*, London: Harper Collins.

Gibson-Graham, J. D. (2008) "Diverse Economies: Performative Practices for 'Other Worlds'", *Progress in Human Geography*, vol 32, no 5, pp. 613–632.

Glaser, J. (2015) "Microplastics in the Environment", *Clean Technologies and Environmental Policy*, vol 17, no 6, pp. 1383–1391.

Glorieux, I., et al. (2010) "In Search of the Harried Leisure Class in Contemporary Society: Time-use Surveys and Patterns of Leisure Time Consumption", *Journal of Consumer Policy*, vol 33, pp. 163–181.

Goodwin, N. R. (1997) "Overview Essay", in N. R. Goodwin, F. Ackerman and D. Kiron (eds) *The Consumer Society*, Washington, DC: Island Press.

Gregson, N. and Crewe, L. (2003) *Second-Hand Cultures,* Oxford: Berg.

Gregson, N., et al. (2013) "Moving Up the Waste Hierarchy: Car Boot Sales, Reuse Exchange and the Challenges of Consumer Culture to Waste Prevention", *Resources, Conservation & Recycling*, vol 77, pp. 97–107.

Guiltnan, J. (2009) "Creative Destruction and Destructive Creations: Environmental Ethics and Planned Obsolescence", *Journal of Business Ethics*, vol 89, pp. 19–28.

Guiot, D. and Roux, D. (2010) "A Second-Hand Shoppers' Motivation Scale: Antecedents, Consequences, and Implications for Retailers", *Journal of Retailing*, vol 86, no 4, pp. 355–371.

ISWA (2015) *Global Waste Management Outlook*. Ed. Wilson, J. C. et al., http://unep.org/ietc/Portals/136/Publications/Waste%20Management/GWMO%20report/GWMO_report.pdf, accessed 1 December 2015.

Janssen, M. A., Kohler, T. A. and Scheffer, M. (2003) "Sunk-Cost Effects and Vulnerability to Collapse in Ancient Societies", *Current Anthropology*, vol 44, no 5, pp. 722–728.

Kelly, T. (2004) "Sunk Costs, Rationality and Acting for the Sake of the Past", *Nous*, vol 38, no 1, pp. 60–85.

Kennicott, P. (2017) "Hamburg's Shimmering Ship of the Skies", *The Guardian Weekly*, vol 196, no 26 (2 June 2017), p. 40.

Kim, S. and Paulos, E. (2011) "Practices in the Creative Reuse of e-Waste", *Human Factors in Computing Systems: Proceedings of the SIGCHI Conference* (CHI '11), pp. 2395–2404.

Kinna, R. (2006) "William Morris and the Problem of Englishness", *European Journal of Political Theory*, vol 5, pp. 85–98.

Kirk, A.G. (2007) *The Whole Earth Catalog and American Environmentalism*, Lawrence, KS: Kansas University Press.

Kroen, S. (2004) "A Political History of the Consumer", *Historical Journal*, vol 47, no 3, pp. 709–736.

Leone, M. (2015) "Longing for the Past: A Semiotic Reading of the Role of Nostalgia in Present-Day Consumption Trends", *Social Semiotics*, vol 25, no 1, pp. 1–15.

Li, X. (2013) "A Comparative Analysis of Hybrid Car Advertisements in the USA and China: Desire, Globalization, and Environment", *Environmental Communication*, vol 7, no 4, pp. 512–528.

Liboiron, M. (2013a) "Modern Waste as Strategy", *Lo Squaderno: Explorations in Space and Society*, vol 29, pp. 9–12, at www.losquaderno.professionaldreamers.net/?cat=162, accessed 1 May 2015.

Liboiron, M. (2013b) "Plasticizers: A Twenty-first Century Miasma", in J. Gabrys, G. Hawkins and M. Michael (eds) *Accumulation: The Material Politics of Plastics*, London: Routledge.

Lipovetsky, G. (2011) "The Hyperconsumption Society", in K. M. Ekström and B. Glans (eds) *Beyond the Consumption Bubble*, London: Routledge.

Lübbe, H. (2008) "The Contraction of the Present", in H. Rosa and W. B. Scheureman (eds) *High-Speed Society: Social Acceleration, Power, and Modernity*, Philadelphia, PA: Pennsylvania State University Press.

Luckman, S. (2013) "The Aura of the Analogue in a Digital Age: Women's Crafts, Creative Markets and Home-Based Labour After Etsy", *Cultural Studies Review*, vol 19, no 1, pp. 249–270.

Luckman, S. (2015) *Craft and the Creative Economy*, London: Palgrave Macmillan.

MacBride, S. (2013) *Recycling Reconsidered: The Present Failure and Future Promise of Environmental Action in the United States*, Cambridge, MA: MIT Press.

Manzini, E. (2002) "Context-based Wellbeing and the Concept of Regenerative Solution: A Conceptual Framework for Scenario Building and Sustainable Solutions Development", *Journal of Sustainable Product Design*, vol 2, pp. 141–148.

Manzini, E. and Tassinari, V. (2013) "Sustainable Qualities: Powerful Drivers of Social Change," in R. Crocker and S. Lehmann (eds) *Motivating Change: Sustainable Design and Behaviour in the Built Environment*, London: Earthscan/Routledge.

Manzo, J. (2014) "Machines, People, and Social Interaction in 'Third-Wave' Coffeehouses", *Journal of Arts and Humanities*, vol 3, no 8, pp. 1–12.

McNeill, J. R. (2010) "The Biosphere and the Cold War", in M. P. Leffler and O. P. Westad (eds) *The Cambridge History of the Cole War*, 3 vols, Cambridge: Cambridge University Press, vol 3, pp. 422–444.

McNeill, J. R. and Engelke, P. (2014) *The Great Acceleration: An Environmental History of the Anthropocene Since 1945*, Cambridge, MA: Harvard University Press.

Meikle, J. (1997) "Material Doubts: The Consequences of Plastics", *Environmental History*, vol 2, no 3, pp. 278–300.

Meikle, J. (2005) *Design in the USA*, New York: Oxford University Press.

Miller, D. (2008) *The Comfort of Things*, Cambridge: Polity.

Morrison, K. A. and Minnis, J. (2013) *Carscapes: The Motor Car, Architecture, and Landscape in England*, London: Yale University Press.

Parrott, K. R., et al. (2008) "Kitchen Remodelling: Exploring the Dream Kitchen Projects", *Housing and Society*, vol 35, no 2, pp. 25–42.

Paull, J. (2007) "Certified Organic Forests & Timber: the Hippocratic Opportunity", in *Proceedings of the ANZSEE Conference* (Australia New Zealand Society for Ecological Economics), http://orgprints.org/11042/, accessed 1 December 2015.

Petrini, C. (ed.) (2001) *Slow Food: Thoughts on Taste, Tradition and the Honest Pleasures of Food*, Washington, DC: Chelsea Green.

Pfister, C. (2010) "The '1950s Syndrome' and the Transition from a Slow Growing to a Rapid Loss of Global Sustainability", in F. Uekoetter (ed) *Turning Points of Environmental History*, Pittsburgh, PA: University of Pittsburgh Press.

Princen, T. (2002) "Distancing: Consumption and the Severing of Feedback," in T. Princen, M. Maniates and K. Conca (eds) *Confronting Consumption*, Cambridge, MA: MIT Press.

Pugh, T. and Weisl, A. J. (2013) *Medievalisms: Making the Past in the Present*, London: Routledge.

Quin (2016) "How Many Cups Does Starbucks Use in a Day?", Leo Quin, www.leozqin.me/leo-does-the-math-how-many-cups-does-starbucks-use-in-a-day/, accessed 20 March 2017.

Ritivoi, A. D. (2002) *Yesterday's Self: Nostalgia and the Immigrant Identity*, New York: Rowman and Littlefield.

Ritzer, G. and Jurgenson, N. (2010) "Production, Consumption, Prosumption: The Nature of Capitalism in the Age of the Digital 'prosumer'", *Journal of Consumer Culture*, vol 10, no 1, pp. 13–36.

Rollins, W. (2006) "Reflections on a Spare Tire: SUVs and Postmodern Environmental Consciousness", *Environmental History*, vol 11, pp. 684–723.

Rome, A. (2003)"'Give Earth a Chance': The Environmental Movement and the Sixties," *Journal of American History*, vol 90, no 2, pp. 525–554.

Rosa, H. (2003) "Social Acceleration: Ethical and Political Consequences of a Desynchronized High-Speed Society", *Constellations*, vol 10, no 1, pp. 3–33.

Rosa, H. (2010) "Full Speed Burnout: From the Pleasures of the Motorcycle to the Bleakness of the Treadmill: The Dual Face of Social Acceleration", *International Journal of Motorcycle Studies*, vol 6, no 1, http://ijms.nova.edu/Spring2010/IJMS_Artcl.Rosa.html, accessed 1 May 2014.

Rosa, H. (2013) *Social Acceleration: A New Theory of Modernity*, trans. J. Trejo-Mathys, New York: Columbia University Press.

Rosenzweig, R. and Thelen, D. (1998) *The Presence of the Past: Popular Uses of History in American Life*, New York: Columbia University Press.

Sandino, L. (2004) "Here Today, Gone Tomorrow: Transient Materiality in Contemporary Cultural Artefacts", *Journal of Design History*, vol 17, no 3, pp. 283–293.

Sanne, C. (2002) "Willing Consumers – Or locked-in? Policies for a Sustainable Consumption", *Ecological Economics*, vol 42, pp. 273–287.

Sanne, C. (2005) "The Consumption of Our Discontent", *Business Strategy and the Environment*, vol 14, pp. 315–323.

Schiermer, B (2014) "Late-modern Hipsters", *Acta Sociologica*, vol 57, no 2, pp. 167–181.

Schor, J. S. (2005) "Prices and Quantities: Unsustainable Consumption and the Global Economy", *Ecological Economics*, vol 55, no 3, pp. 309–320.

Sennett, R. (1998) *The Corrosion of Character*, New York: Norton.

Sennett, R. (2007) *The Culture of the New Capitalism*, New Haven: Yale University Press.

Seyfang, G. (2009) *The New Economics of Sustainable Consumption: Seeds of Change*, Basingstoke, UK: Palgrave Macmillan.

Skelton, A. and Allwood, J. (2013) "Product Life Trade-offs: What If Products Fail Early?", *Environmental Science & Technology*, vol 47, no 3, pp. 1719–1728.

Slade, G. (2006) *Made to Break: Technology and Obsolescence in America*, Cambridge, MA: Harvard University Press.

Smart, B. (2010) *Consumer Society: Critical Issues and Environmental Consequences*, London: Sage.

Soron, D. (2009) "Driven to Drive: Cars and the Problem of 'Compulsory Consumption'", in J. Conley and A. T. Mclaren (eds) *Car Troubles: Critical Studies of Automobility and Auto-Mobility*, Farnham, UK: Ashgate.

Spoehr, J. (2017) "Stormy Times", in J. Schultz and P. Allington (eds) *Griffith Review 55: State of Hope*, Melbourne: Text Publishing.

Steffen, W., Crutzen, P. J. and McNeill, J. R. (2007) "The Anthropocene: Are Humans Now Overwhelming the Great Forces of Nature?", *Ambio*, vol 36, no 8, pp. 614–621.

Strong, R. C. (1996) *Country Life, 1897–1997: The English Arcadia*, London: Boxtree.

Suri, J. (2010) "Counter-Cultures: The Rebellions against the Cold War Order, 1965–1975", in P. Leffler and O. P. Westad (eds) *The Cambridge History of the Cold War*, 3 vols, Cambridge: Cambridge University Press, vol 3, pp. 460–481.

Tatzel, M. (2014) "Introduction", in M. Tatzel (ed) *Consumption and Well-Being in the Material World*, London: Springer.

Tomlinson, J. (2007) *The Culture of Speed: The Coming of Immediacy*, London: Sage.

Tonsley (2017) https://tonsley.com.au/, accessed 10 February 2017.

Turner, F. (2005) "Where the Counterculture Met the New Economy: The WELL and the Origins of Virtual Community", *Technology and Culture*, vol 46 pp. 485–512.

Turner, F. (2006) *From Counterculture to Cyberculture: Stewart Brand, the Whole Earth Network, and the Rise of Digital Utopianism*, Chicago: University of Chicago Press.

Van Nes, N. (2010) "Understanding Replacement Behaviour and Exploring Design Solutions", in T. Cooper (ed) *Longer Lasting Products: Alternatives to the Throwaway Society*, Farnham, UK: Gower.

Viera, R. A. (2011) "Connecting the New Political History with Recent Theories Temporal Acceleration: Speed, Politics and the Cultural Imagination of Fin de siècle Britain", *History and Theory*, vol 50, pp. 373–389.

Waddesdon (2016) "Waddesdon Manor: History of the House", www.waddesdon.org.uk/house/history-of-the-house, accessed 1 May 2016.

Walker, S. (2011) *The Spirit of Design: Objects, Environment and Meaning*, London: Earthscan.

WAN (2015) "World Architecture News: Adaptive Reuse Award", https://backstage.worldarchitecturenews.com/wanawards/project/tonsley-main-assembly-building-and-pods/?source=categorywinners&mode=gallery&selection=winner, accessed 1 December 2016.

Ward, S. (2012) "Breathing New Life in the Corpse: Upcycling Through Adaptive Reuse", in S. Lehmann and R. Crocker (eds) *Designing for Zero Waste: Consumption, Technologies and the Built Environment*, London: Earthscan.

Wieser, H. (2016) "Beyond Planned Obsolescence: Product Lifespans and the Challenges to a Circular Economy", *Gaia*, vol 25, no 3, pp. 156–160.

Part 1

Culture, meaning, and value

2 Using art to research diverse economies

Social experiments in re-valuing waste

Max Liboiron

Introduction

Modern waste is an economic description. Economies are systems of value; they set the general parameters of what is deemed valuable, what is not, and on what terms. Circulation and exchange are two ways these parameters are enacted – and contested. As an object moves through different economies with different regimes of value, it can become more or less valuable, more or less trash, even though the object itself does not change (Lepawsky and Liboiron 2015). This is how 'one person's trash' can become 'another person's treasure': there are multiple regimes of value. All forms of reusing waste engage in these politics of value at the scale of the individual or business. But what would this revaluation look like as an entire system of value, rather than moments of valuation? How can moments of resistance to consumerism, through reclaiming, reusing, and redesigning trash, scale up?

In 2010, I explored this question through an art exhibit called *Salt-winning: Equal to or Greater Than.* Gallery viewers were invited to take away art made of trash so long as they left something behind of equal or greater value. When participants made an exchange, they filled out surveys outlining how they determined the relative worth of what they took and what they left behind. This chapter analyzes trends in these articulations of value – experimental and fledgling economies – as a way to make regimes of value and their attendant economies apparent and available to politics and action. Cultural theorist Fredric Jameson (1994, p. xii) has argued, "It is easier to imagine the end of the world than it is to imagine the end of capitalism." Environmental apocalypse comes to mind more readily than diverse economies. Yet we can come to see economies that are not premised on constant consumption, disposability, obsolescence, profit, and externalizing pollution 'through' the products of constant consumption, disposability, obsolescence, profit, and externalizing pollution. That is, through the exchange and circulation of repurposed trash, we can 'imagine the end of capitalism', or at least the proliferation of alternatives to capitalism and one of its current strategies, consumerism.

After a short review of key scholarship on the relationship between value and trash, this chapter details the methodologies of *Salt-winning* and its findings. Art can generate data; documenting economic exchange in an art gallery allows "a way of pinpointing the individual and group decisions that influence the unpredictable trajectories of diverse economies," a methodology that geographer of diverse economies J.K. Gibson-Graham (2008, p. 618) argues is key to bringing "marginalised, hidden and alternative economic activities to light in order to make them more real and more credible as objects of policy and activism." Art praxis can generate a 'politics of possibility' (in the spirit of Worsham and Olson 2007), where political work can focus on a different model of business-as-usual so new futures are conceivable.

The value of waste

Under capitalism, waste is a profit-producing strategy. Not only is municipal solid waste a recession-proof multimillion-dollar industry, but the current norms of modern waste and the throw-away society were intentionally cultivated as a money-generating strategy (Liboiron 2013; Slade 2009; Packard 1960). In the 1950s, American industry faced saturating markets in a post-Depression culture where repair, reuse, and the creation of durable objects were common – and moral – practices. Cars, for example, lasted for years, came in only one color, and did not have new versions every year (Slade 2009). Opportunities for growth through the sale of commodities, and thus profit, were rapidly diminishing. The response was garbage. Industry intervened on a material level and developed disposability through planned obsolescence, single-use items, cheap materials, throwaway packaging, fashion, and conspicuous consumption (Packard 1960). These changes were supported by a regimen of advertising that telegraphed industrial principals of value into the social realm, suggesting the difference between durable and disposable, esteemed and taboo, that taught Americans how to waste properly (Strasser 1999). American industry designed a shift in values that circulated goods through, rather than into, the consumer realm. It was a conscious effort to turn consumption into consumer*ism*, an economic order-cum-ideology that promotes the ever-increasing acquisition of goods and services. Recent work by scholars such as Morgan Robertson has shown how this use of waste to create profit has only intensified and become more creative under advanced capitalism as the goal to produce more profits with diminishing resources intensifies (2012). Ecological Marxists have even argued that the current configuration of capitalism 'depends' on creating waste to maintain profit, and this makes the system inherently unsustainable (Horton 1997; O'Connor 1994; Baran and Sweezy 1966). In these theories, waste creates value 'because of' its movement into worthlessness, which creates space for more value via production and consumption.

In her work on the politics of waste in socialist and post-socialist Hungary, Zsuzsa Gille (2007) writes about how capitalism and socialism are not just economies in the colloquial sense, but ways of organizing social life that have repercussions on how waste is created. While both socialist and capitalist economies were "wasteful" in Hungary, she argues that they produced and understood waste differently.

Gille (2010, p. 1056) extends the term "resource regime," which describes how the "specific set of social institutions that determine what natural resources are considered valuable by society, that lay down the principles of valuation, and that resolve the resulting value conflicts" to that of waste. "Waste regimes" are "concerned with the production, circulation, and transformation of waste as a concrete material" (ibid). She argues against theories of waste-society relationships that are exclusively focused on value and skip over materiality, which also has acute social consequences. It matters whether waste is made of apple cores or arsenic. Forgetting the materiality of waste when considering economies of reuse and recirculation misses crucial points for understanding how value is produced, since apple cores and arsenic circulate differently and are routed through different regimes of value.

This insistence on materiality responds to literature in discard studies that defines waste exclusively in terms of value. Richard Thompson's (1979) *Rubbish Theory*, for example, creates a taxonomy of objects that includes transient objects, which are 'used up' or are expected to decrease in value (such as trailers in trailer parks and vegetables), durable objects, which are expected to last and increase in value (such as antiques and art), and trash, which has no value whatsoever. He argues that objects can only move from the transient category (low value) to the durable category (high value) by becoming trash first. His examples include Victorian houses, antiques, and 'retro' fashion. Thompson's definition of waste is based entirely on value, without regard to materiality. Plenty of other things exist in the transient and waste categories that are not taken up as durables, and explaining which re-circulate and which do not has to take account of the materiality of objects, their social context, and the values and meanings that are already salient in a culture. Despite the dematerialization of objects in his work, Thompson set the theoretical stage for one of the mainstays of discard studies: things are not inherently, irreducibly rubbish. Rather, trash 'becomes' trash. Social and cultural organizations and concepts of value play key roles in this becoming.

The 'trash phase' of an object in *Rubbish Theory* can be compared to Arjun Appadurai's (1988, p. 13) analysis of the commodity phase, an object in *The Social Life of Things*:

> Let us approach commodities as things in a certain situation, a situation that can characterize many different kinds of thing, at different points in their social lives. This means looking at the commodity potential of all things rather than searching fruitlessly for the magic distinction between commodities and other sorts of things.

Rather than searching for a 'magic distinction' between waste and not-waste (or commodity and non-commodity), we can focus on the waste potential of things at different points in their social lives. The case study of *Salt-winning* considers the mechanics for how things move in and out of waste and non-waste phases, both in terms of material aesthetics (via trash art) and exchange (via re-valuation).

Salt-winning: a methodology

During December 2009, I gathered trash and road salt in the city of Nelson, British Columbia, Canada. Most of the waste came from dumpsters, and one outside a second-hand store was particularly fruitful. These items were twice discarded: first by their original owners and again by the staff that sorted saleable from unsaleable objects for the store. I also traveled to the local landfill to collect empty glass jars, which I then slumped in a glass kiln to create hunchbacked bell jars. Dumpsters also provided ready-made glass vitrines in the form of light fixtures, broken vases, and orphaned coffee crafts. I used the waste to build miniature dioramas. I then placed the diorama in a water and road salt solution to let the salt water wick up the objects and evaporate, leaving behind a crystalline salt crust. The finished product was sealed in a bell jar.

Figure 2.1   One of 127 trash-art objects available for exchange in *Salt-winning*. Max Liboiron. Untitled. 2010. Mixed media. Private collection. 3″ x 6″.

Photo: Max Liboiron.

The result was designed to look valuable; the salt gave a white, somewhat mysterious crystalline appearance to objects, and objects placed under bell jars have long been associated with value, whether in scientific, design, or museum cultures. At the same time, the objects were reused (repurposed) rather than recycled (ground up and used as raw stock), so their origins as broken toys, Christmas tree ornaments, and pickle jars were still apparent. The art was designed to look clean and precious without hiding the appropriated nature of the materials, an aesthetic I assumed would make the items desirable but clearly previously discarded, and thus lead to complications determining their value.

In January 2010, 127 of these recovered, salted, glass encased objects were displayed at Oxygen Art Gallery. Visitors could take any piece so long as they left behind something of equal or greater value. If they completed an exchange, they filled out a survey that detailed what they took, what they left behind, how they determined the comparative value of the two objects, and what they intended to do with the object. The instructions for the exchange were written on the wall of the gallery, and a gallery attendant was always on hand to facilitate participation. The wall text also made it clear that the objects were made of trash. It read:

Salt-winning:

a: *the deliberate production of salt from seawater*
b: *the extraction of a valuable substance from worthless muck*
    Everything in this exhibition is made from local garbage, salt salvaged from winter streets, and discarded glass.
    You can take away any piece at any time, provided that you leave something of equal or greater value behind in its place.
    Please fill out a survey sheet if you exchange anything.
    Traded items remaining at the end of the exhibition will become property of the artist and may be trashed or donated to groups that have helped make this exhibition possible through their generosity of time, space, money, and garbage.

In total, 82 objects were exchanged and 80 surveys were completed. The anonymous surveys were pinned to the wall at the back of the gallery for others to read.

Salt-winning was a social experiment where latent, internalized concepts of value and popular ideas about waste could be made manifest. Appadurai (1988, p. 5) writes that,

> Thus, even though from a theoretical point of view human actors encode things with significance, from a methodological point of view it is the things-in-motion that illuminate their human and social context.

This art-exhibit-come-economic-experiment queried things-in-motion and the premises that prompted their circulation. The waste-phase of each art object was still apparent, but its current prestige as an art object put it in a liminal moment in its life. This was not a necessarily permanent transition from trash to art, as any objects left over at the end of the exhibit were returned to the waste stream. There was no guarantee that any exchanges would occur, or that exchanges would be valued at more than trash itself. In fact, during a call-in radio interview about the exhibit before it opened, a caller said he had a pair of old sneakers he had been meaning to take down to the Salvation Army he would bring for the exchange. He was proposing an exchange between two low-worth items only valuable to someone of a lower economic status. I was concerned this would become a trend during the exhibit. Participants had every right to value the art as trash. This would have been legitimate data.

Salt-winning is what Appadurai (1988, p. 21) calls a "tournament of value," where results were not assured:

> Tournaments of value are complex periodic events that are removed in some culturally well-defined way from the routines of economic life. [. . .] Finally, what is at issue in such tournaments is not just status, rank, fame, or reputation of actors, but the disposition of the central tokens of value in the society in question. Finally, though such tournaments of value occur in special times and places, their forms and outcomes are always consequential for the more mundane realities of power and value in ordinary life.

Appadurai refers to formal events such as fairs, festivals, and rituals, where status and rank are contested through objects. Historian Michael Conan (2002), for example, writes about how garden art during and just after the Renaissance was central to defining both individual and family standing as part of bourgeois and aristocrat classes, but as amateurs and lower class groups began to cultivate gardens, gardens also redefined social hierarchies and the boundaries of elite classes. *Salt-winning* focuses on the latter aspect of tournaments of value. Rather than looking to status and fame, it focuses on how "central tokens of value in [a] society" in an elite art gallery setting – namely, trash – can be used to contest value in "the more mundane realities of power and value in ordinary life" (Appadurai 1988, p. 21). For Appadurai, a token of value, whether they are gardens, trash, or commodities, is based on performances of social relationships with the thing. Key performances during tournaments of value can disrupt and redefine not only individual social relations between people, but also the standing of the object and wider practices of valuing objects.

The type of tournament *Salt-winning* engenders is closely related to Gille's (2010, p. 1056) concept of a waste regime, where trash-art is being used to consider what is valuable, and how to "lay down the principles of valuation" when the exchange is outside of normal economic transactions. These principals and logics of valuation – how things are valued and why – are being contested as much as the art's categorization as trash. The tournament determines whether or not the trash-art moves out of the trash phase, but more importantly for this research, 'how' that trash-art is valued and whether or not individual decisions coalesce into systems of value – economies – rather than mere expressions of preference via consumption that have consequences for the more mundane realities of power and value in ordinary life.

A typology of value

Local radio, newspapers, gallery advertisements, and word of mouth about *Salt-winning* focused on the raw materials of the art – trash – and how gallery visitors could take the trash-art away. In personal conversations and during participant observation, people spoke about how they expected the art to look like trash. Upon entering the gallery, they were often surprised by the crystalline beauty of the objects and a genuine desire to own one. One woman came to the exhibit opening without something to exchange, since she assumed she would not be participating. The first hour of the opening was just for looking, so visitors could see the full range of objects. The moment the art was available for exchange the woman tore off her jewelry in exchange for a small diorama with a burned-out glass light bulb inside. On her survey, she stated that the jewelry was not valuable enough, and she would return with more objects to make it an equal exchange. The next day, there was a deer skull wearing her jewelry. She had completed her exchange.

Many exchanges had stories like this one. People realized they had little practice articulating and engaging in economic equivalence. Sometimes people had to create something they felt was equivalent, so they made art or other objects with specific points of comparison to the piece they wished to take. The following typology groups these varied behaviors and narratives together based on the survey question: "How did you determine that the object you left behind was of equal or greater worth than the object you took?" Many surveys included more than one measure of equivalence, so the chart count exceeds 80.

Most responses cluster loosely into two dominant categories: affective value – using feelings, experiences, and sentiment to adjudicate or transfer value – and value created by labor and materials through the creation of similar objects,

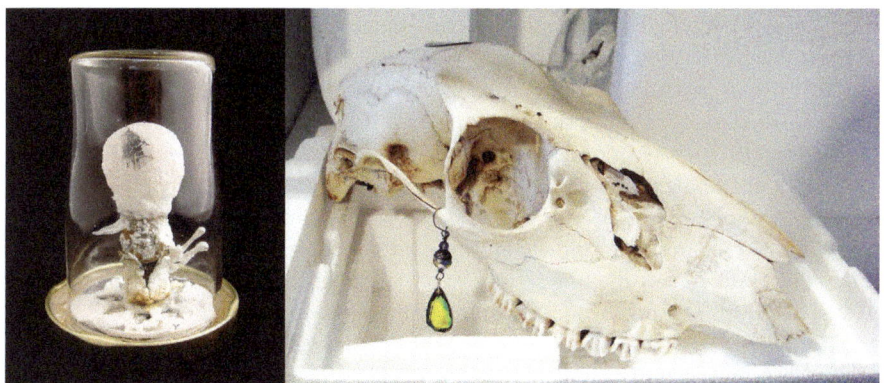

Figure 2.2 Two items exchanged in *Salt-winning*

Photo: Max Liboiron.

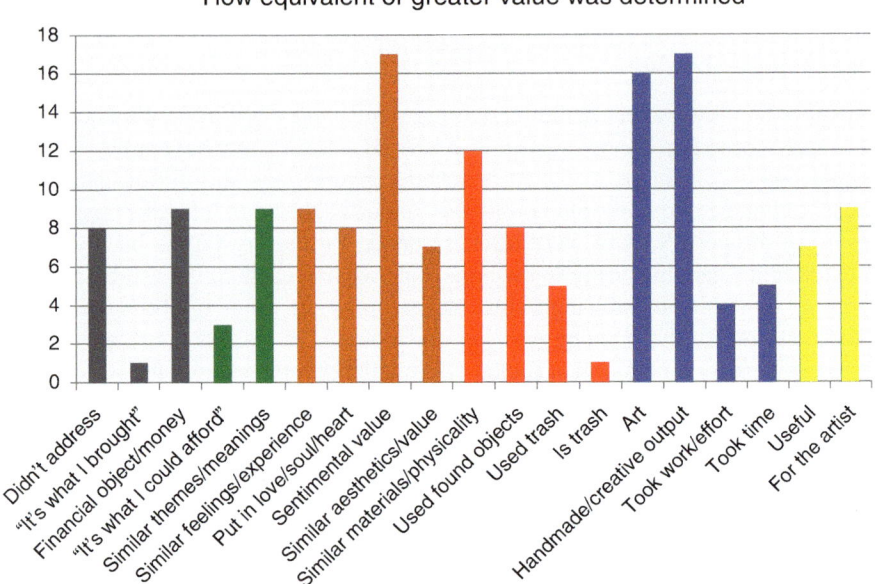

Figure 2.3 Number of times a *Salt-winning* survey mentioned a type of value

usually handmade objects, often with found materials. There were also genres of valuation that were less common. All modes of valuation will be explored in detail here, no matter how often they arose.

Self-maximization and profit

Most participants had little or no practice determining equivalent value. In the dominant everyday economic system, we seek out deals, bargains, and sales to pay as little as possible for something, or we are trying to make a profit and so look to sell things for more than they cost us to produce. These are normal economic activities. Determining exactly what something is worth, and then finding an equivalent tender, is not. This system is legitimized by popular truisms that humans are inherently greedy, wasteful, or will ensure the tragedy of the commons, and so the best and only economic system that will work must be governed by assured mutual self-maximization. Economist Adam Smith (Smith and Dickey 1993 [1910], p. 11), for example, argued that people must "address ourselves, not to [a shopkeeper's] humanity but to their self-love, and never talk to them of our own necessities but of their advantages." Even though Smith's invisible hand, the tragedy of the commons, and the idea that humans are inherently wasteful or greedy have been disproved (Piketty 2014; Ostrom et al. 1999; Appell 1993), they remain strong cultural narratives used to explain, encourage, and excuse self-maximizing behaviors and practices. Challenging these narratives are, in effect, the symbolic and cultural stakes of a project like *Salt-winning*.

Of the 82 exchanges, six might be considered self-maximizing. Seven percent of exchanges left behind something determined as less valuable, allowing the participant to gain more than he or she contributed. For example, a segment of copper pipe was left behind. The accompanying survey stated: "It's what I brought." In four instances, the surveys made the case that the exchanged objects were equivalent, but there were acute discrepancies: one of the largest, most intricate pieces in the exhibit was exchanged for a small, mass-produced sticker, for example. For most of these cases, there is no explicit indication that the participant thought their exchange accrued more value than they gave, but my own comparison of the objects, as well as discussions by other gallery visitors who read the surveys, show a discrepancy. Yet, even counting these grey instances, profit-seeking exchanges were the rarest type of transaction in *Salt-winning*.

Monetary value and moral economies

Eleven people left money or financial tokens. In addition to a lottery ticket, a student discount card (with $1.01 in change), and a gift certificate (worth $50), eight people left cash. Five people left $50. There were also single instances of $40, $100, and $300. The person who left $300 also left $100 for a second exchange (a much smaller piece) on a subsequent visit. Just under half of the surveys for monetary exchanges did not address how equivalence was determined (including the person who left the $100 and $300), but for those that did, most cited it was "what I could afford" and noted that money is useful, particularly for making more art.

One participant who left $50 and a signed first edition of a book written by a local author, wrote:

> I was going to leave the book alone, because it connects to this place. It felt inadequate so I added what money I could afford, hoping it will be useful.

Another person who left $50 wrote,

> I'm not sure if it was equal or greater considering the artists work [*sic*]. It was what I could afford and I wanted the piece.

It is notable that exchanges consistently clustered at $50. This is what anthropologist E.P. Thompson (1971) calls a moral economy. Based on the peasant movements of eighteenth century England, Thompson's example of a moral economy is one where starving villagers, rather than bakers, set the price of bread and other necessary foodstuffs. They enforced the standard through riots if prices rose or scarce items were sold to buyers outside the local area. During these direct actions, "men and women of the crowd were informed by a belief that they were defending traditional rights or customs; and, in general, that they were supported by the wider consensus of the community" (Thompson 1971, p. 78). Prices were set according to what villagers needed and could afford. This was seen as the way to set the proper price of food; any deviations were not only seen as a financial burden, but, more importantly, immoral. In *Salt-winning*, $50 was the price set for exchanges by the community, for the community. While there were neither riots nor a shared socioeconomic class among participants, making an exchange for below $50 would have appeared self-maximizing – and thus immoral – given the accumulation of exchanges at $50 or higher. Indeed, the only amounts of financial tender for less than $50 were not cash, but lottery tickets and savings cards where the quantitative value was not as visible or was less determined. Even the gift certificate discussed below, which had no outward indication of how much money is was worth, was for $50.

Other surveys showed that making exchanges within a moral economy is more complicated than merely stating a common price. In the first moments of exchange during the opening, a couple quickly obtained a large and much-coveted piece in exchange for a "hand turned wooden goblet filled with sea salt and a large rolled [*sic*] amethyst with two blue daises." But they changed their mind. They wrote:

> Though I/we had originally chosen the large sea anemone in a coffee pot, we loved it but felt guilty (for it was larger [than the piece we eventually took]). We traded down. There is something empowering about trading and about dictating value based on internal compass [*sic*].

After they put the "large sea anemone in a coffee pot" back, it was immediately exchanged by another participant for $300.

Sentimental value, sacrifice economies, and affective equivalence

Sentimental value was one of the highest types of value participants used to determine equivalency between objects. Sentimental value occurs "if and only if the thing is valuable for its own sake in virtue of a subset of its relational properties" that may include connections to family, friends, and significant personal experiences (Fletcher 2009, p. 56). Sentimental objects left behind in *Salt-winning* included: a "recent letter from a one-time lover"; "the CD of one of my dearest friends"; and a necklace worn by the participant's late best friend. I will focus on two elements of a sentimental economy. First, the non-transferable nature of sentimental value, and second, how participants maneuvered quantifying sentimental value so it might be equal to or greater than another object.

Philosopher Anthony Hatzimoysis (2003, p. 374) notes that while sentimental value is personal, it is "in principle intelligible by everyone else, even though [it] may not be applicable to anyone else. Universality may thus come in through the understanding that something has the relevant [sentimental] quality, but not in the experience of that quality." That is, the recipient of a sentimental object can understand, but not feel, sentimental value. This non-transferability makes it an interesting choice for exchange, since the object will likely always be of higher value to the donor. One survey even remarked that the way the participant determined that the object they exchanged was equal or greater value was because the object held memories for them. Memories, especially if not explained, are not transferable.

Thus, one of the most commonly used sources of value could not be fixed, transferred, nor fully appraised. Yet, sentimental economies accounted for some of the most extreme accounts of value in the exhibit. For example, one participant cried during her exchange. She had left:

> My lucky pouch. It contains 2 very special stones; one from my best friend in the whole world and one from my summer lover. Both of the people play a huge role in my life and my creativity and happiness. Wearing the pouch makes me feel really calm and lucky. I think it's important to share the luck and be able to share it with someone who's sharing their art with me.

I was at the gallery during this exchange, and the young woman told me her best friend had died the previous year. The rock was one of the few tangible objects she had left.

This exchange created what I call a sacrifice economy, where the value of the object left behind was of significantly more value than a new object to the point that the exchange produced pain. Other sacrifice economies exist where more value is put into a system than recovered – parenting and some forms of volunteering come to mind, as well as unpaid household labor that accounts for up to 50 percent of economic activity in both rich and poor countries (Ironmonger 1996) – but those tend to be systems where the sacrifice directly benefits another person or group of people. In *Salt-winning*, there was no such transfer of benefits. More than one participant spoke about the cathartic nature of parting with sentimental objects in a semi-public forum, but even if this is a benefit, it remains solidly on the giving side of the exchange. In a sentimental economy, the flow of value is constantly interrupted.

However, several survey respondents speculated how equivalence, and even the transference of value, 'could' occur in a sentimental economy. Some participants indicated that the degree of discomfort of parting with a sentimental object matched their desire for the new art object. This was also a technique when feelings were not necessarily sentimental. Someone who left behind four small magnets depicting his or her paintings wrote,

> I went with the feeling I had when I saw this particular piece and choose paintings that created a similar feeling for me in their making. Value = feeling (rather than value = $) [*sic*].

They used affective equivalence to determine value equivalence. Affect is the manifestation of emotion. Three participants endeavored to make affect transferable.

> Well, it's about sensation. Looking at ART really make me feel happy and warmer in my chest [*sic*]. Just like a pair of mittens could warm you up.

Participants left a pair of mittens, pickled asparagus, and a gift certificate to an organic food market with the intention of having the artist and recipient of exchanged objects feel the same way the participants felt – warm, satiated, full. They were chosen to create equivalent and transferable affect.

Synonyms

The tally chart of how equivalent or greater value was determined shows that many participants aligned aesthetics, materials, physical attributes, or concepts between the object they took and the one they left behind. If the two objects for exchange were similar sizes, colors, materials, or dealt with similar themes, the idea was that they were also equivalent in value, or at least ideal for exchange. In many cases, people used found materials, trash, glass, and/or salt to build objects for exchange, or traded art for art. Equivalence became synonymy. This would result in an odd economy if scaled up to a more systemic level; we would be trading linen for linen and coats for coats, rather than linen for coats and vice versa.

Labor theory of value

The labor theory of value, popularly articulated by economist Karl Marx (1867), states that the value of an object created for exchange is conferred by the amount of labor necessary to produce it; the greater the labor, the greater the value; and the more specialized the labor, the greater the value. In the words of one participant: "Because they were both things we created, I figured it a fair trade." Others wrote explicitly about the time or effort they put into making a piece for exchange, and the equivalence of labor type – handcrafting – was a common explanation of equal value. Specialized labor was rarely remarked upon, except that professional artists made some exchanges to another professional artist, but a drawing made by a five-year-old was also determined equivalent, so specialization and professionalization did not factor into all labor-based exchanges.

Use theory of value

For Marx (1867), use value is the want-satisfying power of a good or service. Many surveys described their object's general, abstract usefulness for general, abstract wants: "Hopefully it's useful." Occasionally, a participant indicated that the object was specifically useful to the recipient – me. However, the use theory of value doesn't necessarily describe an alternative economy. Marx argues that because use-value is the basis for the creation and circulation of many goods, regardless of whether they are commodities designed for trade, that use value tells us little about the specific character of the economy in which they are produced and sold.

Persisting exchanges and the temporality of value

Some of the most interesting responses to *Salt-winning* were by people who returned numerous times to tweak their exchange, seeking to get the equivalence just right. One person left a book whose main theme reminded her of the piece she took. She returned a few days later to add an essay she had written inspired by the trade. She then returned once more to swap out the original book with a first edition copy of the same book. On another occasion, an artist invited me to his studio so I could choose a piece I wanted. I chose a ceramic sculpture. Several days later, two more had joined it at the gallery. Then two more. In the end, a set of five ceramics completed the exchange rather than the single one I had chosen myself. The corresponding survey read: "Added 2 more of same because 3 was incomplete." The artist spoke to me about concerns he had that as he grew to like his object more over time, his exchange would become less equal and he would have to keep updating his exchange, even after the exhibit was over. In these instances, economic exchange is not a single transaction but an ongoing relationship between objects, people, and shifting values. An economy based on such relations would be a densely networked, ongoing set of local interpersonal and inter-object relations, where single transactions mean something different than they do in a supermarket today.

Conclusion: diverse economies and a politics of possibility

Considerations of economic systems and value are a mainstay of discard studies (the study of waste) across disciplines, from Vance Packard's (1960) classic *The Waste Makers* to the more recent *The Business of Waste* (Stokes et al. 2013). Both historically and in the present day, much time, money, and ink has been spent on thinking about how to re-value the detritus of production and consumption. Yet, in almost all cases, these works are about critiquing present systems or tweaking the status quo to promote efficiency and extract more of the same kind of value. Most of this work sees capitalism and one of its main economic strategies, consumerism, as large, powerful, and intractably indivisible behemoth systems. Diverse economies is a term developed in the pioneering work of J.K. Gibson-Graham (2008, 2006a, 2006b) to destabilize the axiomatic conflation of 'the economy' with 'capitalism' and its attendant modes of exchange with consumerism. *Salt-winning* is part of this same kind of research (see North, 2015 for an excellent review of diverse economies), which seeks to document the massive variety of non-capitalist economic relations that exist within capitalism as well as outside of it – and even in spite of it.

Such work, whether textual or artistic, aims to move away from using capitalism and consumerism as the categories that matter most to understand systems of value, waste, consumption, and exchange, particularly when looking for alternatives. In *Rule of Experts*, economic geographer Timothy Mitchell (2002, p. 248) considers how to approach historical agricultural land reforms in Egypt and their shifting systems of value:

> Can one find a way to take local complexity and variation and make it challenge the narrative of the market? Can one do so without positing the existence of a precapitalist or noncapitalist sphere, or even multiple capitalisms, positions that always reinvoke the universal nature of capitalism

[or consumerism]? To begin to do so, we have to stop asking whether rural Egypt is capitalist or not. We have to avoid the assumption that capitalism has an 'is' and take more seriously the variations, disruptions, and dislocations that make each appearance of capitalism, despite the plans of the reformers, something different.

Mitchell argues that the economy is not a monolithic, stable, transcendental force, but a series of performative projects that have been naturalized through economic policy and practices of accounting. *Salt-winning* is a performative platform that changes such accounting. The variation in valuation it engenders allows us to disentangle value from profit, waste from valuelessness, and consumerism from consumption. This is not just so we can describe and critique economic problems with more nuance (allowing us to say, for example, that it is consumer*ism* rather than aggregated consumption per se, at the crux of the issue), but also let's see how other practices of consumption, waste, and exchange might continue in different economies, but in fundamentally different ways (see, for example, Lepawsky and Liboiron 2015, Liboiron 2015).

Salt-winning is a project in a larger cross-disciplinary research community that makes elements of value and diverse economies visible and viable. At the core of this research is the desire to scale up from individual actions (of reuse, revaluation, design, and un-trashing) to systems. In isolation, turning trash into art is a form of resistance against consumerism, disposability, and capitalism. It is a "weapon of the weak" (Scott 1985) based on "foot-dragging, evasion, false compliance, pilfering" (xvi) and individual acts of recovery and tinkering. As Samantha MacBride (2011) has pointed out in *Recycling Reconsidered*, reuse efforts are not scaling up. No matter how large a lot of reclaimed, reused, and refurbished objects arc for sale in the wider economy, they pale in comparison to the production, circulation, and disposal of new goods to the point that their impact remains neglient in monetary terms, but, more importantly, in systemic terms. Rather than tweak the system or engage in isolated acts of resistance, research on diverse economies looks to change the political terrain more broadly, including how things are valued and exchanged.

Our goal is to make diverse economies manifest and available to the imagination, and thus to politics and action. *Salt-winning* is about providing a space to "enact and construct rather than resist (or succumb to) economic realities" (Gibson-Graham 2008, p. 7). Reused items culled from the trash are ideal objects on which to found this enactment because their value is so easily contested, and enabling a wide range of variations, disruptions, and dislocations that make each appearance of consumption and exchange something different.

References

Appadurai, A. (1988) *The Social Life of Things: Commodities in Cultural Perspective*, Cambridge: Cambridge University Press.

Appell, G. N. (1993) *Hardin's Myth of the Commons: The Tragedy of Conceptual Confusions,* Working Paper No. 8, Phillips, ME: Social Transformation and Adaptation Research Institute.

Baran, P. and Sweezy, P. (1966) "Monopoly Capitalism", in *An Essay on the American Economic and Social Order*, New York: Monthly Review Press.

Conan, M. (2002) *Bourgeois and Aristocratic Cultural Encounters in Garden Art, 1550–1850*, Cambridge, MA: Harvard University Press.

Fletcher, G. (2009) "Sentimental Value", *The Journal of Value Inquiry*, vol 43, no 1, pp. 55–65.

Gibson-Graham. J. K. (2006a) *Postcapitalist Politics*, Minneapolis, MN: University of Minnesota Press.

Gibson-Graham. J. K. (2006b) *The End of Capitalism (As We Knew It): A Feminist Critique of Political Economy*, Minneapolis, MN: University of Minnesota Press.

Gibson-Graham, J. K. (2008) "Diverse Economies: Performative Practices for Other Worlds", *Progress in Human Geography*, vol 32, no 5, pp. 613–632.

Gille, Z. (2007) *From the Cult of Waste to the Trash Heap of History: The Politics of Waste in Socialist and Postsocialist Hungary*, Bloomington, IN: Indiana University Press.

Gille, Z. (2010) "Reassembling the Macrosocial: Modes of Production, Actor Networks and Waste Regimes", *Environment and Planning A*, vol 42, pp. 1049–1064.

Hatzimoysis, A. (2003) "Sentimental Value", *The Philosophical Quarterly*, vol 53, no 212, pp. 373–379.

Horton, S. (1997) "Value, Waste and the Built Environment: A Marxian Analysis", *Capitalism Nature Socialism*, vol 8, no 2, pp. 127–139.

Ironmonger, S. (1996) "Counting Outputs, Capital Inputs and Caring Labor: Estimating Gross Household Output", *Feminist Economics*, vol 2, pp. 37–64.

Jameson, F. (1994) *The Seeds of Time*, New York, NY: Columbia University Press.

Lepawsky, J. and Liboiron, M. (2015) "Why Discards, Diverse Economies, and Degrowth?", *Society & Space Open Forum*, http://societyandspace.com/material/discussion-forum/discards-diverse-economies-and-degrowth-forum-by-josh-lepawsky-and-max-liboiron/why-discards-diverse-economies-and-degrowth-josh-leapwsky-and-max-liboiron/, accessed 15 October 2016.

Liboiron, M. (2013) "Modern Waste as Strategy", *Lo Squaderno: Explorations in Space and Society*, vol 29, pp. 9–12.

Liboiron, M. (2015) "An Ethics of Surplus and the Right to Waste", *Society & Space Open Forum*, http://societyandspace.com/material/discussion-forum/discards-diverse-economies-and-degrowth-forum-by-josh-lepawsky-and-max-liboiron/an-ethics-of-surplus-max-liboiron/, accessed 15 October 2016.

MacBride, S. (2011) *Recycling Reconsidered: The Present Failure and Future Promise of Environmental Action in the United States*, Cambridge, MA: MIT Press.

Marx, K. (1867) *Capital, Volume I*, Harmondsworth: Penguin/New Left Review.

Mitchell, T. (2002) *Rule of Experts: Egypt, Techno-Politics, Modernity*, Berkeley, CA: University of California Press.

North, P. (2015) "The Business of the Anthropocene? Substantivist and Diverse Economies Perspectives on SME Engagement in Local Low Carbon Transitions", *Progress in Human Geography*, vol 40, no 4, pp. 437–454.

O'Connor, M. (1994) *Is Capitalism Sustainable?: Political Economy and the Politics of Ecology*, New York: Guilford Press.

Ostrom, E., Burger, J., Field, C. B., Norgaard, R. B. and Policansky, D. (1999) "Revisiting the Commons: Local Lessons, Global Challenges", *Science*, vol 284, pp. 278–282.

Packard, V. (1960) *The Waste Makers*, Philadelphia, PA: D. McKay Co.

Piketty, T. (2014) *Capital in the Twenty-First Century*, Cambridge, MA: Harvard University Press.

Robertson, M. (2012) "Measurement and Alienation: Making a World of Ecosystem Services", *Transactions of the Institute of British Geographers,* vol 37, no 3, pp. 386–401.

Scott, J. (1985) *Weapons of the Weak: Everyday Forms of Peasant Resistance*, New Haven: Yale University Press.

Slade, G. (2009) *Made to Break: Technology and Obsolescence in America*, Cambridge, MA: Harvard University Press.

Smith, A. and Dickey, L. (1993 [1910]) *Wealth of Nations (Abridged)*, Cambridge, MA: Hackett Publishing.

Stokes, R. G., Köster, R. and Sambrook, S. C. (2013) *The Business of Waste: Great Britain and Germany, 1945 to the Present*, Cambridge: Cambridge University Press.

Strasser, S. (1999) *Waste and Want: A Social History of Trash*, New York: Palgrave Macmillan.

Thompson, E. P. (1971) "The Moral Economy of the English Crowd in the Eighteenth Century", *Past & Present*, vol 50, no 76–136.

Thompson, M. (1979) *Rubbish Theory: The Creation and Destruction of Value*, New York: Oxford University Press.

Worsham, L. and Olson, G. A. (2007) "The Politics of Possibility: Encountering the Radical Imagination", http://philpapers.org/rec/WORTPO, accessed 14 March 2015.

3 Repurposing cultural heritage collections

The aesthetics and meaning of reuse

Sally Butler

Introduction: the cultural heritage of waste

> Behind the scenes at museums the world over are thousands upon thousands of artefacts. These objects, removed from other lives, other places, other times, are neatly labelled, catalogued and packed away out of sight, rarely displayed and infrequently studied.
>
> (Byrne et al. 2011, p. 3)

The many thousands of objects referred to in the above excerpt from *Unpacking the Collection: Networks of Material and Social Agency in the Museum* [henceforth *Unpacking the Collection*] are the result of the twentieth century's rampant institutional consumption of cultural heritage material (Belk 1995; Byrne et al. 2011). Collecting is a model form of consumption whether on private (Belk 1995; Martin 1999) or institutional levels (Bal 1994; Baudrillard 1994 [1968]; Macdonald 2006). Institutions such as the Smithsonian, the Louvre and most national museums in the Western world, amassed such vast collections that they conventionally display less than 3 percent of their holdings at any given time. Budgets are straining under the weight of collection storage, documentation, and research expenses, just as the demands of a growing cultural tourism industry push for more diversity and innovation in exhibition and public programs. Waste in this context of cultural heritage material goes far beyond lost opportunities occasioned by the limited visibility of an overabundance of objects. The real wealth of these collections is, of course, about knowledge rather than objects, and understanding the cultural creativity that gives meaning and shape to artifacts from the past and present. It is ironic that the Western cult of collecting and cataloging that developed out of a post-Enlightenment thirst for knowledge should have lost its way in privileging an accumulation of objects that has virtually buried its own treasure. Twentieth century collection and cataloging practices developed into a self-referential system that effectively distanced the object from the social relations and geographical environments that generated their original purpose (Battiste 2000; Byrne et al. 2011; Christen 2011), and the hallowed vaults of collection storage threatened to become a wasteland of lost wisdom.

Unpacking the Collection articulates how current museological practice is addressing this problem by placing greater emphasis on repurposing the collection in support of the more future-orientated aims of cultural sustainability and regeneration. The once static perception of a 'collection' is transforming into a notion of a dynamic tool where collection holdings are put to use and reuse, where they are circulated, contested, repurposed, and re-valued. Leading curators and scholars in the field of cultural heritage who contributed to this volume, and others like it (Karp and Levine 1991; Elsner and Cardinal 1994; Barringer and Flynn 1998; Edwards et al. 2006), articulated a new energy in museums to make collections 'active' and to increase the accessibility and circulation of their material. Byrne et al. (2011, p. 3) describe the global momentum to "turn the museum itself into an active field site where existing knowledge structures are contested and regenerated." Emphasis here on 'structures' of knowledge defers to greater concern for epistemological aesthetics and awareness of the 'conditions' of knowledge (Hughes 2007). In other words, 'structures' of knowledge, 'shapes' of time, and cultural 'forms' privilege aesthetic as distinct from logocentric frameworks in how we make sense of the world, and how this aesthetic of reuse is culturally coded. In a broad sense, there is a discernible shift towards engaging more overtly with a cultural imagination driven by creative shapes of knowledge, time, and space.

Indigenous 'source communities' whose previous generations produced the majority of material held in cultural heritage collections are instrumental in driving change in museological practice and in advancing recognition of culturally diverse knowledge paradigms (Alpers 1991; Lakoff and Johnson 1999; Classen and Howes 2006; Dei and Kempf 2006; Kawano 2011). Postcolonial politics of the 1980s and 1990s created global momentum in the assertion of the rights of the colonized to self-determination regarding their own past, present, and future, and this in turn impacted museum practices of collection, documentation, interpretation, and display. Museums now regularly consult with source communities regarding collected cultural artifacts and engage in debates about issues such as repatriation, re-attribution, and restricted access (Peers and Brown 2003; Kurin 2004; Gazi 2014). Transformations are also apparent within a more fundamental system of order, such as critique of how museum collections implicitly shape (and stereotype) knowledge through generalized categories such as the Primitive, Asian, Oceanic, etc. (Price 1989, 2007).

The Convention for the Safeguarding of Intangible Cultural Heritage (ICH) ratified by the United Nations Educational Scientific and Cultural Organization (UNESCO) in 2003 represents formal global affirmation of these new museological practices in repurposing cultural heritage collections towards a stronger emphasis on cultural sustainability (UNESCO 2003; Bortolotto 2007). It embodies a rhetoric of creativity and continuity that supports the new spirit of reuse in cultural heritage management:

> Intangible Cultural Heritage means the practices, representations, expressions, knowledge, skills – as well as the instruments, objects, artefacts and cultural spaces associated therewith – that communities, groups and, in some

cases, individuals recognise as part of their cultural heritage. This intangible cultural heritage, transmitted from generation to generation, is constantly recreated by communities and groups in response to their environment, their interaction with nature and their history, and provides them with a sense of identity and continuity, thus promoting respect for cultural diversity and human creativity. For the purposes of this Convention, consideration will be given solely to such intangible cultural heritage as is compatible with existing international human rights instruments, as well as with the requirements of mutual respect among communities, groups and individuals, and of sustainable development.

(UNESCO; Article 2 – Definitions)

UNESCO's affirmation of the intangible also reflects the radical temporal reorientation in cultural heritage politics (Kurin 2004; Hoelscher 2006). It encourages governments to alter a mindset for preserving the past in a static form of temporal isolation towards a creative reuse of the past that incurs dynamic interactions between the past, present, and future. These creative interactions of time constitute a climate of thought where unused objects and archives held in cultural heritage collections constitute an act of waste built on a history of over-consumption and resultant devaluing of the cultural sustainability inherent to what we mean by 'heritage'.

Repurposing collections and producing the past

The shift in emphasis from 'building' collections to 'unpacking' collections is also nurtured by Michel Foucault's studies into the power relations embedded in archives and how the past is ordered, staged, and used (Foucault 1970, 1972). New museology adheres to a politics of resistance against Western hegemony (Battiste and Henderson 2000; Dei and Kempf 2006; Kawano 2011) that gained momentum in the late twentieth century when texts such as Foucault's (1970) *The Order of Things* deepened insight about the dynamics of social power based on the determinations and ordering of knowledge. In the wake of these ideas a global community of difference now identifies itself as a co-presence of diverse knowledge paradigms, or what Foucault (1970, p. 168) calls 'epistemes', involving conditions of non-linear historicity and synchronicity. In other words, the concept of a master narrative no longer guides the historic measure of time, and instead we engage with a flux of diverse narratives that overlap, conflict, and loop backwards and forwards through time. Foucault's later work regarding the nature of the episteme engenders a quite radical interpenetration of the past, present, and future that models a cycle of time as opposed to a linear advance of time.

In *The Order of Things*, Foucault (1970) proposed an episteme to signify the historical background behind an epoch's structures of knowledge. This inherited epistemological construct, or system, conditions how we make sense of the world. However, in his later *Power/Knowledge* publication Foucault (1980) reconsiders the role of an episteme as something more temporally variable. He rejects the singular grounding of an a priori epistemological paradigm and argues instead for the simultaneous co-existence of different epistemes. This means that within a given epoch different historical constructs of knowledge operate at the same time and 'interact'. The idea of multiple simultaneous epistemes impacts significantly on how the past is conventionally 'distanced' from the present in museum displays. Foucault's effort to unshackle lived experience of time from a singular chronological construct of time participated in a broader philosophical movement in the twentieth century (involving such figures as Henri Bergson, Martin Heidegger, and Jean François Lyotard) to rethink relationships between the past, present, and future.

Susan Crane's (2006) contextualization of Foucault's approach to time in terms of heritage museums argues that the latter create an illusion of timelessness, as though visitors have stepped back in time. The illusion creates a diminished awareness of the discrepancy between past and present fueled by visitors' desire to apprehend (or consume) the past beyond the passage of time. However, it is significant to identify that Crane is referring to an 'illusion of time' here, generated by the heritage industry's trade in desire and consumption of the past. Multiple and simultaneous epistemes described by Foucault (above) are far from timeless and are not an illusion. They are the frameworks of time that shape how we know the world today and involve an acute awareness of multiple and overlapping timeframes. Rather than anesthetizing our distinctions of time, non-linear historicity and synchronicity create a highly active engagement with how yesterdays (and tomorrows) are actively produced today. This escalated proximity between what is past, present, and to come, facilitates an enhanced register of accountability in our relationships with the past and future. It is arguably an accelerated cultural imagination that moves as swiftly backwards, and sideways, as it does forward.

There is obviously much more at stake in unpacking and repurposing museum collections than a decolonizing principle of cultural difference, and we are yet to adequately comprehend what kind of cultural capital ensues from museums' initiatives to repurpose wasted areas of collections into active fields of contestation, accessibility, and circulation. This 'active' field is, however, undoubtedly more creative and inherently involves a greater emphasis on aesthetics. Creativity, imagination, and aesthetics play a significant role in endowing the past with a 'presence' – an active character of the past less alienated from the present. In fact, the French theorist Jacques Rancière (2004) describes the entire modern (and postmodern) era as a distinctly "aesthetic regime" that derives principally from a new relationship with the past. Rancière's (2004, pp. 12–45) concepts of political aesthetics, and "the distribution of the sensible" in particular, are useful for understanding the politics of visibility, accessibility, and circulation invested in "unpacking collections" and an aesthetics of reuse.

Rancière's (2004) theory of political aesthetics considers the infrastructure of power involved in how we negotiate the sensible world, and how an epoch's "distribution of the sensible" determines what is seen and unseen, audible and inaudible, sayable and unsayable. This approach construes a politics invested in how objects and forms circulate and orchestrate public consciousness, construing the aesthetics of circulation and access as instruments of power. This is not simply a politics of representation so much as a concept of a sensory collective 'unconscious' that develops its own vocabulary of response to what stimulates attention. In a chapter titled "The Distribution of the Sensible: Politics and Aesthetics," Rancière (2004, pp. 12–30) extends Foucault's thesis regarding power engendered in hierarchies of knowledge to consider these hierarchies aesthetically. Aesthetics in this instance refers to how something presents itself to sense experience and broadens Foucault's concerns for how something presents itself through language. Rancière (2004) argues that living in the sensible world involves an implicit ordering of perception that encompasses the 'totality' of the visible and invisible, audible and inaudible, sayable and unsayable. Politics here is deeply invested in not only what gets said, but also involves structures of the sayable, visible and audible, concomitant with the shapelessness (or 'waste') of what gets ignored (or in the case of museum collections, what gets stored or hidden).

Cultural heritage collections in this context are instruments of cognitive orientation determined by the "perceptual coordinates of a community" (Rockhill 2004, p. 3). Rancière (2004, p. 9) argues that art and its capacity to redistribute or recalibrate sensible forms has a vital role in political life, where the "aesthetic acts as configurations of experience (that) create new modes of sense perception and induce novel forms of political subjectivity." These aesthetic coordinates of a community impact significantly on the politics of the past and how the past is valued. Cultural artifacts *qua* art held in museum collections become potential configurations of experience and forms of political subjectivity when they are 'unpacked' and reused in the active field of museum collections. Rancière does not specifically address the politics of exhibition and display as part of the aesthetic activities determining the "distribution of the sensible", but museums and art galleries are obviously a powerful influence on public taste and modes of perception and, in turn, respond to fluctuating demands of public appeal and expectations. These collections are also increasingly subject to a global ethics of display (Gazi 2014; Stark 2011), overseen by global institutions and agencies such as UNESCO. More 'active' museum collections of the present implicitly adopt a significant role in mediating the "distribution of the sensible", and inevitably register, and impact upon, changes in not only what we know and value about the past, but 'how' we know and value the past.

Concepts of time are crucial in determining patterns of production, circulation, and consumption. Time is money, but it also creates other value – cultural, social, and even religious (in respect of mortality, etc.). The dynamics of how the 'sensible' is distributed is, for Rancière, fundamentally a matter of time and how a given epoch relates to the past. He argues that today's distinctly aesthetic turn derives from modernity's "two regimes of historicity" (2004, p. 24). Modernity is cast as an "aesthetic regime" along the familiar (Romantic) lines of an era where aesthetics is liberate from its function as (mimetic) representation and retreats from everyday life into a liberated realm of creative imagination. Such ruptures with the past are embedded in notions of modernist avant-gardism – the aesthetic and political qualities of what it is to be modern. However, Rancière (2004, p. 24) also contends that this is a simplistic notion of modernity's relationship with the past, and he proposes instead a more contradictory temporality. Modernity's desire for rupture with the past also (paradoxically) "incessantly restages the past." Museums factor into modernity's 'distribution of the sensible' at this point:

> The aesthetic regime of the arts invents its revolutions on the basis of the same idea that caused it to invent the museum and art history, the notion of classicism and new forms of reproduction . . . And it devotes itself to the invention of new forms of life on the basis of an idea of what art *was*, an idea of what *would have been.*
>
> (Rancière 2004, p. 25, emphasis in original text)

This is not the picture of a radical rupture with the past but instead a modernity that reuses the past, reshapes it, and repurposes it. We could say that modernity even 'advances' the past. This is a mindset not dissimilar to overlapping temporalities in familiar sayings, such as "he lives in the past" or "she lives for the future." Rancière's (2004) argument is that these shapes of time, and relationships with time, are also a powerful factor in our collective cultural imagination and fundamentally impact decisions about truth, knowledge, and value. UNESCO's (2003) Convention for the Sustaining of Intangible Cultural Heritage is an example of a charter for reuse of the past in line with the perceptual coordinates of today's global community.

Conservative concepts of museums as institutions of conservation and preservation – a role that maintains historical distance with the past – fail to take into account modernity's twin relationship with the past that simultaneously alienates and 'incessantly' restages the past through its cultural institutions and disciplines of knowledge. Museums' current momentum in unpacking collections, developing active fields of engagement, and reconnecting collections with the contemporary cultural practices of source communities heed the reuse of the past that Rancière argues always underpinned what has defined modernity. The "rarely displayed and infrequently studied" artifacts referred to in *Unpacking the Collection* manifest as an immense scale of institutional overconsumption and underscore, following Marx (1956 [1867]), how the exchange value of heritage commodities outgrew their use-value. The real value of artifacts accumulated (consumed) by modernity's museum collections lay in their regenerative potential for "the invention of sensible and material structures of a life to come" (Rancière 2004, p. 29).

Art that currently repurposes cultural heritage collections captures this potential as an aesthetic of reuse.

Torres strait island art repurposing the Haddon collection

> It is important in our time to not only rely on the old stories, but also use this time to invent new things for our generation and the next ones. We need to invent new objects and artefacts and we have the technology to make things easier. . .
>
> (Nona 2005, p. 29)

Visual art is one of today's most prominent methods in repurposing cultural heritage collections. Aesthetic engagement with the past involves perceptual encounters between the past, present, and future, and this is nowhere more evident than with the contemporary art movement that emerged in the early 1990s in Australia's Torres Strait Islands (TSI). Contemporary TSI art is highly diverse but remains cohesive within the aesthetics of reuse underpinning their genre of linocut printmaking. The printmaking expresses a cultural and political solidarity that galvanized national and international attention to the continuity and vitality of TSI culture. TSI's print movement is also significant in terms of how it repurposes the "material forms and structures of knowledge" of the Haddon Collection in the Cambridge University Museum of Archaeology and Anthropology (CUMAA), one of the oldest and most comprehensive cultural heritage collections in the world. As the foregoing quote by one of TSI's most prominent artists implies, this new aesthetic in TSI art recalibrates the 'distribution of the sensible' and instills a new order of the visible, audible, and sayable regarding TSI culture today, and its relationship to the past and future.

Until the 1990s, TSI art could be described as a victim of the 'distribution of the sensible' because it was virtually invisible. The print movement that emerged in the 1990s responded to a state of virtual invisibility fueled by a complex historical background and isolated geographic location. TSI is located at the crossroads between the Indian and Pacific oceans and between the Asian and Australian continents. It is close to Melanesia and in very close proximity to Australia's Cape York Peninsula to the south and Papua New Guinea to the north. Surrounding sea channels have seen centuries of maritime traffic from across the world and experienced considerable Japanese immigration after the pearling industry commenced in the 1860s (Herle and Philp 1998; Mitchell 2008; Mosby 2011). Christian Missions established during the nineteenth century significantly eroded traditional cultural practices, a situation exacerbated in the twentieth and twenty-first centuries by a large proportion of the TSI population moving to the Australian mainland in search of employment and educational opportunities.

Despite being at a crossroads, the TSI remained almost unknown to the Australian public until 1992 when Eddie Mabo, an elder from the Torres Strait's Mer Island, succeeded in gaining land title (Native Title) for his homeland in a landmark decision in the High Court of Australia (*Mabo versus the Queensland Government* 1992). This decision rejected the doctrine of 'terra nullius' that substantiated the Commonwealth of Australia's ownership of land. Australia's Indigenous land rights movement began decades earlier but no prior decision had overturned the *terra nullius* principle of "land belonging to no one." The British claim on Australian land had been based on the false premise that the Indigenous population lacked any form of land ownership. Three years after the Australian Federal Government passed the Native Title Act, the Wik people from Aurukun in North Queensland succeeded in gaining recognition of coexisting title with pastoral leases, extending the significance of the Native Title decision. This hub of Indigenous political activity across northeastern Australia stimulated an arts movement that celebrated their communities' newfound recognition and reclaimed rights. For the TSI, reclaiming cultural traditions from the past represented a method of consolidating political solidarity that would help sustain the cultural sustainability of its geographically dispersed population.

Numerous TSI artists involved in the printmaking movement expressed their relationship to the past by 'restaging' and 'repurposing' a breadth of material held in the Haddon Collection. Alfred C. Haddon had created this collection in conjunction with his work as a British marine biologist turned anthropologist and ethnologist. His most notable research involved fieldwork in the TSI in 1888/9, followed by a year-long expedition in 1898 (Haddon 1903; Herle and Philp 1998; Mosby 2011). The 250 objects collected by Haddon in 1888/9 entered the collection of the British Museum, but the 1898 expedition resulted in an immense collection of 1,300 objects covering ethnographic material, photographic documentation, and wax cylinder recordings that are now held principally in CUMAA (Herle 1998). Haddon conducted extensive interviews while in the TSI and was assisted in translations by John Bruce, a resident teacher working with the Anglican Mission. Bruce was an enthusiastic amateur ethnologist and also assisted Haddon with follow-up research, documentation, acquisitions, and commissions for several years after the completion of the expedition. Over the following century, the Haddon Collection drifted in and out of circulation depending on fluctuating attitudes to the study of cultural heritage (another example of the fluctuating 'distribution of the sensible'). The Collection gained unprecedented circulation when the Expeditions' results were published in several volumes titled *The Reports on the Cambridge Expedition to the Torres Straits* in 1903. By the 1930s a new regime of visibility diminished the Haddon Collection into obscurity. Anita Herle (1998, p. 101) describes how the growing popularity of field research, as opposed to collection research, meant that ethnographic collections were rarely brought into the presence of scholarly attention. Visibility improved by the 1980s when the spirit of 'unpacking collections' brought the Haddon material back into focus, and by the 1990s many of the TSI community took an active interest in the Haddon Collection, particularly in terms of its potential role in cultural sustainability.

Herle was CUMAA's Senior Curator of World Anthropology during the transition of the Haddon Collection from obscurity to a new regime of community outreach activities (including a CUMAA exhibition celebrating the centenary of the Haddon collection in 1998 [Herle and Philip 1998]) and regards the Collection today as an active mediator between the past and the present. Herle describes the expeditions' results as "an important resource for the emergence of a strong contemporary Islander identity. Turtle-shell masks, feathered headdresses, drums and models make important intercultural links between past and present, between Europe and the Islands" (1998, p. 78).

TSI artists themselves also acknowledge the significance of the Haddon collection for contemporary cultural expression. Alick Tipoti, one of the leaders of the linocut print movement, recalled:

> Something stirred inside me when I first saw the prints from Alfred C. Haddon's report and Myths and Legends of the Torres Strait by Margaret Lawrie.

The lecturers [at his art college] would have copies of these on display as examples to assist us in our work, and little did they know how it inspired me to be where I am today.

(Islander Magazine 2011)

The Margaret Lawrie Collection mentioned by Tipoti did not contain the ethnographic nor visual detail of the Haddon Collection, but did augment it by creating records of the continuity of TSI intangible cultural heritage into the present (Lawrie 1970). From 1964 to 1973, Lawrie documented a living oral tradition in storytelling and mythmaking, and took extensive notes regarding TSI genealogies and the modern life of women (Mosby 2011). These stories and legends were illustrated by a number of TSI artists of the time, and the narrative basis of these illustrations is a precursor to the 'heritage reuse' aesthetic of the subsequent 1990s printmaking movement. Lawrie's Collection is now held in the State Library of Queensland's John Oxley Library and includes notes from her field research that were used as evidence of cultural continuity in the aforementioned *Mabo versus the Queensland Government* Native Title case of 1992 (Mosby 2011, p. 133). The intangible cultural heritage significance of the Margaret Lawrie Collection was also acknowledged in 2008 when it was included as part of the UNESCO Australian Memory of the World Register. These foregoing events contributed to cultural heritage collections becoming 'active', and creating new regimes of visibility fostering cultural sustainability.

TSI printmaking artists gained inspiration from a much broader pool than the Haddon and Lawrie archives of course, particularly in terms of artists consulting with traditional elders and recalling designs, stories, and cultural practices, from their childhood experiences. However, the key elements of the printmaking aesthetic can be found in Haddon's accumulation of illustrations and objects. Five TSI artists and one cultural elder visited CUMAA in 2001 to view the Haddon material (Mitchell 2008). Beverley Mitchell describes what might be called the "multiple historicity" involved in this artistic encounter with a cultural heritage collection:

Seeing such an extensive collection first hand was enormously inspirational for the young artists and gave them direct access to traditional carving designs from the nineteenth century. These artists have also familiarised themselves with Haddon's report on his expeditions to varying degrees, as well as other source books on cultural material, thereby creating a situation where an invested meaning in their choice of pattern is informed by both oral tradition and through the filter of a colonial text.

(Mitchell 2008, p. 6)

We begin to see here how a complex overlapping of "multiple historicity" plays out in TSI contemporary art practice. Filters of time are at work – be they colonial texts, continuous oral traditions handed down to later texts, or the creative imagination of artists wanting to reshape the past for the future. The aesthetics of reuse developed by contemporary TSI printmakers is distinctive and coherent in the manner that it captures this fluidity of timescapes in densely patterned compositions. Forms overlap, merge, and seem to disappear, and patterns once carved into wood and turtle shell are brought into a new order of thought with the modern printmaking technique (see Figure 3.1). The new aesthetic does not copy imagery from the past, but restages it as a fluid matrix of continuity, groundlessness, and ambiguous time. Objects come in and out of visibility with techniques of optical illusions that are similar to how Ludwig Wittgenstein referenced the rabbit or duck example of two ways of seeing exactly the same image. Objects literally move in and out of view as perception adjusts to the conditions of seeing.

Production of linocut printmaking referencing material from the Haddon Collection stimulated a dynamic circulation of knowledge and creativity in the community more broadly. Artists were required to consult with elders, and elders in turn molded their knowledge and memories from the past to produce new knowledge that would be of use to young artists. Consultation involved discussion of all matters of life such as: genealogies, customary land ownership, mythological narratives, and how community morals are embedded in a worldview and practices in material culture, hunting, and caring for the environment. Artists not only produced visual art, but also created dance, song, storytelling, and music in both live and recorded forms. The artists' aesthetic manipulations and reuse of a past that was in part harvested from the Haddon Collection effectively adjusted the perceptual coordinates of how TSI is seen, heard, and said.

The reordering of the senses involved in TSI contemporary art is deeply steeped in complex concepts of time, and the "multiple historicity" at play in this accelerated cultural imagination is apparent in other genres of the TSI art movement. Brian Robinson's art is part of the TSI printmaking movement, but his work also

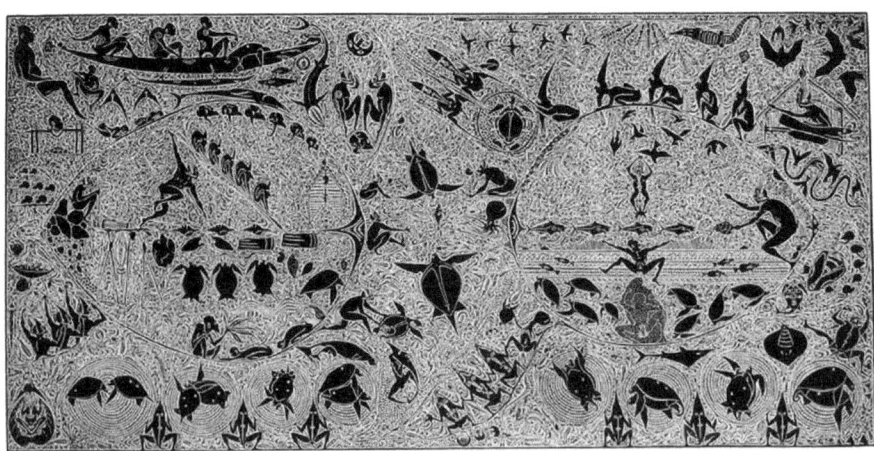

Figure 3.1 Alick Tipoti, "Wadth, Zigin Ar Kusikus" (2006), Linocut print, ed. 50 112 x 211 cm

includes sculpture, public art, and figurative etchings. One particular etching from 2012 epitomizes the time travel involved in TSI aesthetic engagement with the Haddon Collection.

The title of this etching (see Figure 3.2) includes anachronistic reference to a USB flash drive within what appears to be a quote from the original Haddon record. Its lengthy title is: "August 23, 1898 – *Today I collected with much zeal, through the barter and exchange of gifts, ancient artefacts belonging to a race of Indigenous Australians known as Torres Strait Islanders. Wooden masks, pearl shell pendants, smoking pipes, dance objects, and a strange device called a USB flash drive, were among the items obtained. A. C. Haddon*" (Robinson 2012, pp. 28–29). Robinson's image looks down on a table of objects that apparently belong to the Haddon Collection and are rendered with sepia tones giving an anti-quated sense of time. The scattered array includes masks, dance apparatus, and body adornments along with the anthropologist's tools of trade – labels, pencils, and a set of calipers. A USB Flash drive sits at the bottom of the image but does not appear out of place – or out of time (even though the title ostensibly dates the image at 1898). The USB does not immediately come to notice and its anachro-nism is only apparent when the viewer aligns their visual perception with chrono-logical thinking. In this context, the USB signifies a fascinating artifact of the future. Robinson plays with a virtual reality in this image that releases the cultural heritage material from its historical duties and restages them as a contemporary 'memory device'.

Conceptions of the past as a fluid or interchangeable memory device recur throughout contemporary TSI art. Janet Fieldhouse's 2014 exhibition titled *Mark and Memory* captured this approach to time travel beautifully, but in the entirely different medium of ceramics (Fieldhouse 2014). The work in this exhibition

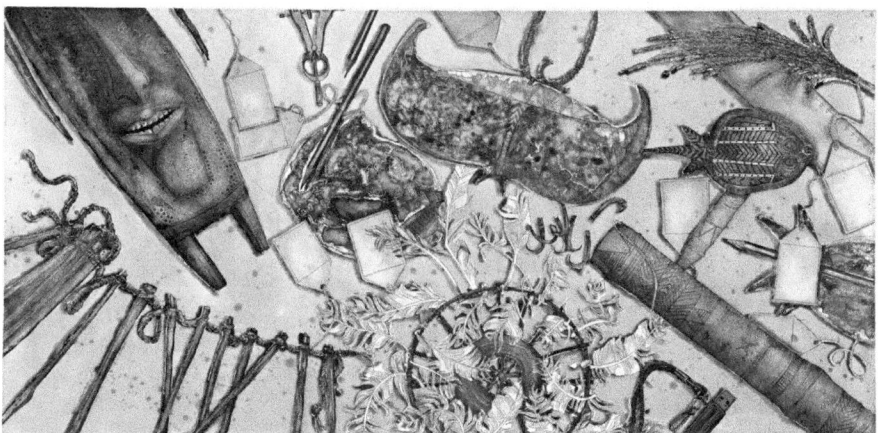

Figure 3.2 Brian Robinson, "August 23, 1898" (2012), Etching printed in three colors from one plate, ed. 30, 495 x 980 mm

Copyright courtesy of the artist and KickArts Contemporary Arts, Cairns repurposed the Haddon collection through the artist's preferred medium of flexible porcelain – a medium that is a perfect metaphor for Fieldhouse's aim to create fluid memories (see Figure 3.3). Flexible porcelain has an organic binding matrix added to standard ceramics raw materials that gives a greater pliability prior to firing and increases the medium's tolerance for creative manipulation. Similar to many artists from the TSI printmaking movement, Fieldhouse completed a Diploma in Visual Arts at Cairns TAFE, and then undertook postgraduate study in Visual Arts at the Australia National University. Her research focused on

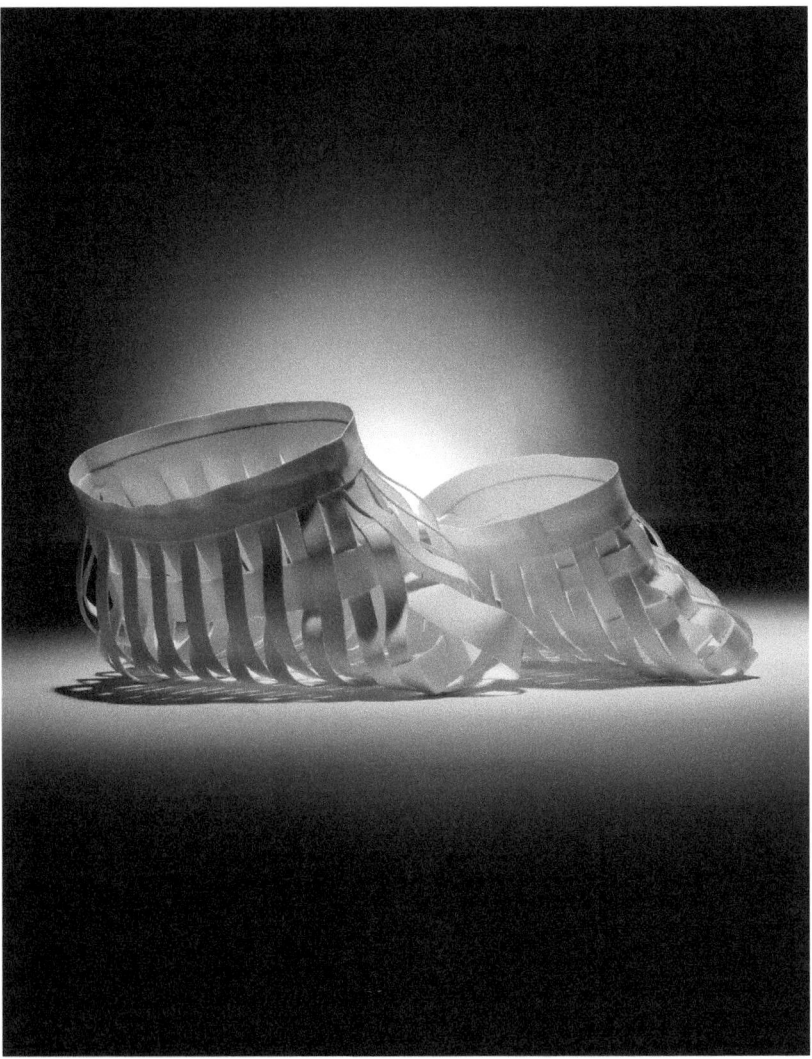

Figure 3.3 Janet Fieldhouse, Memory Series 1 (2014), Porcelain, 30 × 20 × 29 cm

Courtesy of the artist and Vivien Anderson Art Gallery, Melbourne translating the weaving practices of her traditional homeland in TSI's Erub Island into ceramic arts, and subsequently turned to a tradition in women's body scarification (an inkless tattooing technique) that is no longer practiced but is recorded in detail in Haddon documentation. Fieldhouse followed in the steps of her peers by discussing traditional practices with community elders, researching the Haddon archives, and then viewing the collection first hand at CUMAA.

Concepts of time underpinning Fieldhouse's (2014) body of work in *Mark and Memory* are overlaid and fluid. The artist describes her technique in these terms: "the hand created the forms, the memory is the idea created." Memory, the past, is 'created' and re-staged. This manipulation of memory is significant because Fieldhouse is dealing with a shared cultural heritage in her work that is regarded by her TSI community as fundamental to cultural sustainability and political viability. It is no small matter for the TSI community when their artists work creatively with the past, present, and future in their art. Fieldhouse's chosen medium of flexible porcelain is particularly effective in teasing out the actual act of transforming the "distribution of the sensible". Her ceramic forms that represent traditional woven armbands and women's material culture portray this in how forms seem to be in a state of collapse and sag against the grain of their natural state (see Figure 3.4).

It is a similar effect to Salvador's Dali's surrealist dream-state where familiar objects such as clocks and tables appear to melt and lose their shape. Fieldhouse presents unexpected forms that disrupt the objects' fundamental state of being. Similar to Dali's meltdown into the subconscious, Fieldhouse registers this reuse of the past as a restaging of time. Memories, heritage and the past slowly fold into the creative act of cultural regeneration.

Figure 3.4 Janet Fieldhouse, Installation view of Mark and Memory exhibition (2014), Cairns Regional Gallery, Cairns

Fieldhouse's 2014 exhibition also included red raku earthenware objects that appeared strangely suspended between the past, present, and future. Their 'reuse aesthetic' recalls objects seen in the Haddon Collection but are overlaid with a surrealist futurity – they are now a heritage yet to come. Crudely outlined sketches in Haddon's early twentieth century record of TSI material culture take on three-dimensional form in these raku pieces and are decorated with tattoo symbols in white charcoal and delicate cassowary feathers secured by string body-bands. Fieldhouse's Bridge Pendant Series feature white raku shapes modeled on traditional spherical and crescent-shaped symbols – the full and new moon cycle of life. Pendants hang suspended in mid-air in the manner that a memory comes to us – an isolated object of an idea at first, but then starting to tell its own story through detail and associative composition. These red and white raku forms emerge solid and concrete from the raku firing process, but retain the soft-edged suggestiveness of Dalí's time-molten landscapes.

Subverting consumerism

If global consumption continually escalates the rapid production and promotion of the 'new', then we need concerted institutional efforts that counter this momentum by sustaining the 'old' on multiple platforms of continuing relevance. Diane Barthel-Bouchier's 2012 study, *Cultural heritage and the challenge of sustainability*, describes an increasingly organized global heritage community motivated by cultural and environmental sustainability. Sustainability is now the global mantra and is inevitably implicated in debates about consumption – sustainability in this context of course means sustainable consumption (Barthel-Bouchier 2012). The aesthetics of reuse involved in TSI art contributes to debates about sustainable consumption in the manner that the concept of culture is always implicated in multiple timeframes of production and consumption. In the TSI aesthetics of reuse the looping of time implies closer relationships between past, present, and future, and instills a greater sense of accountability between them. A period of time is always implicated in what comes before and after. Consumption cannot occur out of time from the TSI perspective – it can never be unaccountable.

Aesthetics plays with our minds – it works on psychological levels that are subliminal, although powerful, in everyday life. We can see this with the 'salvage aesthetics' of ethnographic exhibitions of the past. Their exoticized distant 'look', or what Rancière would call their 'regime of visibility', derived from a psychology of panic about modernity's rapid rate of change. Fear of change produced a desire for an anchored distant past – pushing the past away just as it was attempting to 'salvage' it. In her analysis of Haddon's mindset as a collector, Anita Herle (1998) foregrounds a similar condition best described as an academic form of panic about a disappearing past and how to 'salvage' the remains. Herle associates Haddon with the anthropological term of "salvage ethnography" that seeks to record the remains of cultural traditions threatened with extinction after the impact of modernity. If there is any validity to the concept of an accelerated cultural imagination, then we perhaps begin to feel the pump of the pedal here in the late nineteenth century with this figure of Haddon. The rupture with the past that modernity 'inscribed' in defining itself, and gave birth to disciplines such as anthropology, also created a psychology of loss that wanted to travel back in time and 'capture' a life before its last breath. This is Rancière's (2004) "multiple historicity" at work – one wanting to resuscitate and the other wanting to take its first new breath.

Herle (1998, p. 77) also refers to Haddon's description of TSI patterns and objects as having "life-histories." Haddon's approach to anthropology drew on his background in biology where he understood patterns and artifacts as objects in a similar fashion to natural specimens. The Darwinian framework popular at the time positioned these objects within an evolutionary process and ecological system where their attributes encoded a genetic inheritance – a 'life-history'. It is a fascinating approach to visual forms in that an ecological history is implanted in the way things look, but from Rancière's (2004) perspective it is an even more fascinating example of perceptual coordinates. The 'life-histories' methodology for looking at visual form implicitly sets up relationships with the past that make it (genetically!) accessible to the present, even if under threat of extinction. TSI artists reactivated these life-histories in developing their reuse art.

The point is that Haddon felt impelled to capture the past rapidly, and in doing so he implemented strategies for restaging the past. One of his most interesting efforts in this regard is his commissioning of replicas and models of artifacts that were no longer in use or impossible to transport (Herle 1998, pp. 87–96). Herle (1998) recounts how this commissioning process set up its own momentum where the TSI community reinvented present practices themselves based on this act of similitude. The entire process of commissioning replicas represents a tension for Herle in terms of Haddon's dualistic Romantic/scientific mindset, a tension apparent in Haddon's description of commissioning Malu masks. Haddon commissioned a reenactment of the Malu ceremony where these masks are worn, and his poetic account of viewing the spectacle drove Herle (1998, p. 92) to write:

> Given the centrality of Malu to the Eastern Islands, how does one reconcile this narrative with Haddon's scientific approach, characterised by strict procedures for collection and verification? This staged reconstruction is not simply the inevitable outcome of a desperate attempt at salvage ethnography, but a shifting, a repositioning which places emphasis on performance and experience.

Desperation about the past impels Haddon to creatively reuse it in the present, and, in the process, TSI participants engaged in cultural regeneration that sustained their sense of identity at a time of crisis incurred by the impact of Christian Missions. Another evolution of this process takes place when TSI contemporary artists re-stage the already re-staged replicas held in the Haddon collection, to again regenerate their cultural traditions at a time of crisis in the modern context of political marginalization and a growing diaspora.

* * *

Throughout these evolutions of restaging and regeneration, the folds of time condense and the desperation to keep life-histories alive escalates. This is arguably the basis of the accelerated cultural imagination that continues to drive our relationships with the past today. It involves an ongoing cycle of life characterized by a fundamental interpenetration of the past, present, and future. This temporal interpenetration implicates different phases of time with each other, and creates a climate of accountability, about what gets produced and consumed across time.

References

Alpers, S. (1991) "The Museum as a Way of Seeing", in I. Karp and S. Levine (eds) *Exhibiting Cultures: The Poetics and Politics of Museum Display*, Washington, DC: Smithsonian Institution Press.

Bal, M. (1994) "Telling Objects: A Narrative Perspective on Collecting", in J. Elsner and R. Cardinal (eds) *The Cultures of Collecting*, London: Reaktion Books.

Barringer, T. and Flynn, T. (eds) (1998) *Colonialism and the Object: Empire, Material Culture, and the Museum*, London: Routledge.

Barthel-Bouchier, D. (2012) *Cultural Heritage and the Challenge of Sustainability*, London: Routledge.

Battiste, M. and Henderson, Y. (2000) *Protecting Indigenous Knowledge and Heritage: A Global Challenge*, Vancouver: University of British Columbia Press.

Baudrillard, J. 1994 [1968], "The System of Collecting", trans. R. Cardinal, in J. Elsner and R. Cardinal (eds) *The Cultures of Collecting*, London: Reaktion.

Belk, R. (1995) *Collecting in a Consumer Society*, London: Routledge.

Bortolotto, C. (2007) "Objects to Processes: UNESCO's 'Intangible Cultural Heritage'", *Journal of Museum Ethnography*, vol 19, pp. 21–33.

Byrne, S., Clarke, A., Harrison, R. and Torrence, R. (2011) "Networks, Agents and Objects: Frameworks for Unpacking Museum Collections", in Byrne, S. Clarke, A., Harrison, R. and Torrence, R. (eds) *Unpacking the Collection: Networks of Material and Social Agency in the Museum*, New York: Springer.

Christen, K. (2011) "Opening Archives: Respectful Repatriation", *The American Archivist*, vol 74, no 1, pp. 185–210.

Classen, C. and Howes, D. (2006) "The Museum as Sensescape: Western Sensibilities and Indigenous Artefacts", in E. Edwards, C. Gosden and R. Phillips (eds) *Sensible Objects, Colonialism, Museums and Material Culture*, Oxford: Berg.

Crane, S. A. (2006) "The Conundrum of Ephemerality: Time, Memory, and Museums", in S. Macdonald (ed) *A Companion to Museum Studies*, Oxford: Blackwell.

Dei, G. J. S. and Kempf, A. (eds) (2006) *Anti-colonialism and Education: The Politics of Resistance*, Rotterdam: Sense Publishers.

Edwards, E., Gosden, C. and Phillips, R. (eds) (2006) *Sensible Objects, Colonialism, Museums and Material Culture*, Oxford: Berg.

Elsner, J. and Cardinal, R. (eds) (1994) *The Cultures of Collecting*, London: Reaktion Books.

Fieldhouse, J. (2014) *Mark and Memory: Janet Fieldhouse* [exhibition catalogue], Cairns, QLD: Cairns Regional Gallery.

Foucault, M. (1970) *The Order of Things: An Archaeology of the Human Sciences*, trans. R. D. Laing, New York: Pantheon.

Foucault, M. (1972) *The Archaeology of Knowledge and The Discourse on Language*, trans. A. M. Sheridan Smith, New York: Pantheon.

Foucault, M. (1980) *Power/Knowledge: Selected Interviews and Other Writings 1972–1977*. Brighton, Sussex: Harvester.

Gazi, A. (2014) "Exhibition Ethics – An Overview of Major Issues", *Journal of Conservation and Museum Studies*, vol 12, no 1, p. 4, DOI: http://dx.doi.org/10.5334/jcms.1021213.

Haddon, A. (ed) (1903) *Reports of the Cambridge Anthropological Expedition to the Torres Strait*, Cambridge: Cambridge University Press.

Herle, A. (1998) "The Life-Histories of Objects: Collections of the Cambridge Anthropological Expedition to the Torres Strait", in A. Herle and S. Rouse (eds) *Cambridge and the Torres Strait: Centenary Essays on the 1898 Anthropological Expedition*, Cambridge: Cambridge University Press.

Herle, A. and Philp, J. (1998) *Torres Strait Islanders: An Exhibition Marking the Centenary of the 1898 Cambridge Anthropological Expedition*, Cambridge: Cambridge University Museum of Archaeology and Anthropology.

Hoelscher, S. (2006) "Heritage", in S. Macdonald (ed) *A Companion to Museum Studies*, www.blackwellreference.com.ezproxy.library.uq.edu.au/subscriber/tocnode.html?id=g9781405108393_chunk_g978140510839317, accessed 29 March 2016.

Hughes, F. (2007) *Kant's Aesthetic Epistemology*, Edinburgh: Edinburgh University Press.

Islander Magazine (2011) "The Man, The Masks and The Magic", www.islandermag.com.au/The-Man-The-Mask-and-The-Magic.51.0.html, accessed 12 August 2014.

Karp, I. and Levine, S. (eds) (1991) *Exhibiting Cultures: The Poetics and Politics of Museum Display*, Washington, DC: Smithsonian Institution Press.

Kawano, Y. (2011) "Being Part of the Cultural Chain", in N. Wayne, A. Kempf, and M. Simmons (eds) *The Politics of Cultural Knowledge*, Rotterdam: Sense.

Kurin, R. (2004) "Safeguarding Intangible Cultural Heritage in the 2003 UNESCO Convention: A Critical Appraisal", *Museum International*, vol 56, no 1, pp. 66–77.

Lakoff, G. and Johnson, M. (1999) *Philosophy in the Flesh: The Embodied Mind and its Challenge to Western Thought*, New York: Basic Books.

Lawrie, M. (1970) *Myths and Legends of the Torres Strait*, St. Lucia, QLD: University of Queensland Press.

Macdonald, S. (2006) "Collecting Practices", in S. Macdonald (ed) *A Companion to Museum Studies*, Oxford: Blackwell.

Martin, P. (1999) *Popular Collecting and the Everyday Self: The Reinvention of Museums?* London: Leicester University Press.

Marx, K. (1956) [1867] *Capital*, F. Engels (ed), Moscow: Progress Press.

Mitchell, B. (2008) Kulba Yadail, Kaien Wakaine Thamam (old lyrics, new expressions), in B. Missi (ed) *Urapun Kai Buai, Contemporary Lino Prints from the Torres Strait*, Cairns, QLD: KickArts Contemporary Arts.

Mosby, T. (2011) "Welcome Home to Torres Straits Where You Had Left Your Heart and Went Away Empty", in *The Torres Strait Islands*, Brisbane, QLD: Queensland Art Gallery/Gallery of Modern Art.

Nona, D. (2005) *Sesserae, The Works of Dennis Nona*, Brisbane, QLD: Dell Gallery, Queensland College of Art.

Peers, A. K. and Brown, L. (2003) *Museums and Source Communities: A Routledge Reader*, London: Taylor and Francis.

Price, S. (1989) *Primitive Art in Civilized Places*, Chicago: Chicago University Press.

Price, S. (2007) *Paris Primitive, Jacques Chirac's Museum on the Quai Branly*, Chicago: Chicago University Press.

Rancière, J. (2004) *The Politics of Aesthetics*, trans and intro G. Rockhill, London: Continuum.

Robinson, B. (2012) *Men + Gods*, [ex.cat.], Cairns, QLD: KickArts Contemporary Arts.

Rockhill, G. (2004) "Translator's Introduction", in J. Rançiere (ed) *The Politics of Aesthetics*, trans and intro G. Rockhill, London: Continuum.

Stark, C. J. (2011) "The Art of Ethics, Theories and Applications to Museum Practice", in J. Marstine (ed) *The Routledge Companion to Museum Ethics: Redefining Ethics for the Twenty-First Century Museum*, London: Routledge.

UNESCO (2003) "Text for the Convention for the Safeguarding of the Intangible Cultural Heritage", www.unesco.org/culture/ich/index.php?lg=en&pg=00006, accessed 23 January 2015.

4 The devil's horns are made from toilet rolls

Creating costumes and communities from 'junk' objects

Claire Langsford

Introduction

A night elf and a warlock stood outside a videogame store in a suburban shopping mall in South Australia. Despite the heat of the January day, they were wearing full armor, brightly painted in blue and gold. The elf carried a hand-carved bow, and the warlock wore a helmet with horns and flowing plumes. The two female cosplayers – amateur costumers – were dressed as characters from the video game *World of Warcraft* as part of celebrations for the store's tenth birthday.

I knew one of the cosplayers and struck up a conversation about their costumes. I asked what their armor was made out of and they replied that it was primarily constructed from leather and foam. I was particularly impressed by the golden horns on the warlock's helmet that stood up about half a meter from her head. I asked how they were able to support their own weight. She said that the horns functioned as a number of different segments put together and, to my surprise, she explained that these segments were actually made out of toilet rolls.

Cosplay (costume play) is a practice centered upon the assembly and perfor-mance of costumes based on pre-existing character designs. These designs are sourced from globalized popular culture texts – films, television series, Western comics, Japanese manga and anime, and video games. Growing in popularity in the early decades of the twenty-first century, the practice has developed through the globalized exchanges of images, materials, aesthetics, competences, and prac-tices (Winge 2006; Lamerichs 2011; Lunning 2011; Peirson-Smith 2013). Driven by an affinity for the character or its source text, an admiration for the aesthetics of the character design, or by the desire to create a costume that is valued by the community, cosplayers, those who practice cosplay, can spend considerable time, effort, and money in the attempt to recreate character designs in the form of wear-able costumes (Lunning 2011; Okabe 2012). Cosplayers remake popular culture texts into costumes, repurposing globalized imagery for their own personalized, creative works. Cosplayers adopt a parallel approach to material items, breaking them into component materials and working on them to suit their own designs and bodies. As active reusers, cosplayers in Australia provide an insightful case study of an alternative, community-specific framing of reuse. Cosplayers demonstrate a distinctive attitude towards reuse that sits outside of environmental and other ethical discourses. Reuse in cosplay is a skilled activity, a marker of a practitioner's thrift and creativity.

The case study of cosplayers in Australia also demonstrates how community-specific understandings of reuse are promoted and recreated. The view of reuse as a skilled activity endures in the Australian cosplay community because it is deeply embedded in the practice's social relations. Reuse, like other consumption and production activities is situated within the structures, values and networks of social practices (Gregson and Crewe 2003; Crocker 2012). In cosplay communities reuse activities are taught to newcomers, promoted by community leaders, emphasized in the gift-giving of reusable materials between practitioners, and skilled reusers are celebrated as master cosplayers. Cosplayers reuse materials to create costumes, but also to create communities and practitioner identities. The example of reuse activities in cosplay supports a community-centered holistic approach to reuse education initiatives and programs. Reuse activities must be embedded within the social practices of a community if they are to flourish and endure.

Cosplay in Australia is a young, highly dynamic practice. Studies of cosplay in other countries, including Japan (Okabe 2012), Hong Kong (Peirson-Smith 2013), and the United States (Lunning 2011) have identified practitioners as being predominantly aged between 16 and 30 years of age. Surveys of Australian popular convention events suggest that local participants are of a similar age (author interview with Dustin Wilson of AVCon, 2010). As a practice, cosplay in Australia lacks many significant formal structures or organizations. Instead, the practice is situated within a network of sites and events, including popular culture conventions such as Oz Comic-Con and Supanova held in major capital cities, local meet-ups, and photo shoots, as well as online spaces such as websites, forums, blogs, and social networking sites, particularly Facebook.

Data from this chapter is drawn from ethnographic fieldwork conducted by the author for a doctoral thesis in anthropology, exploring the recreation of dynamic material and performance practices within cosplay communities in Australia. The primary method employed was long-term participant-observation, a technique whereby the researcher immerses themselves in the lives of the people they are studying: sharing the same social space and performing the same or similar activities to the members of the culture being studied, as well as recording behaviors and practices observed in the form of written field notes (O'Reilly 2005, p. 84). Other methods included creating video recordings of performances and construction activities and carrying out in-depth semi-structured interviews with key informants. Observing and participating in cosplay activities enabled the author to experience cosplay reuse practices first-hand. Journeying from novice to experienced practitioner, the author learned how to view materials like a cosplayer.

Beyond 'textual poaching': cosplayers and the reuse of material objects

Reuse is fundamental to the practice of cosplay as cosplayers recreate pre-existing textual elements in the form of wearable costumes and performances. The idea of fans as re-users, re-mixers, and re-creators has been long established within fan theory literature (see, for example, Booth 2012; Jenkins 2013). Drawing on de Certeau's (1984) notion of agents as tacticians who respond strategically to environments and structures, Jenkins (2013) argues that fans regularly act as 'textual poachers' by reworking and recreating existing textual material. Fans 'poach' elements such as narrative, setting, character, music, or visual designs from existing source texts and recombine them in the production of new works; films, comics, novels, television programs, and video games are recreated as fan fiction, art, songs, costumes, photography, and video (Jenkins 2013). These actions have been read as a form of resistance to the hegemonic dominance of globalized brands and the power of commercial content producers, as fans are reworking and reinterpreting the meanings of these products to suit their own local, community-specific needs (Jenkins 2013).

Within Australian cosplay communities of practice, this reuse is not only textual but material as cosplayers 'poach' and reuse textual elements – characters, visual imagery, narrative elements – in their costume, performance, and photographic recreations, but also reuse or repurpose 'junk' and mundane material objects.

> We are quite good recyclers at the Australian Costumer's Guild, using bits of what people call junk, and we'll always take a piece of junk and turn it into something really cool.
>
> (Ben, ACG Panel, author's transcript 2011)

While many cross-cultural studies of cosplay have explored cosplayers' textual borrowing and recreation (see, for example, Lunning 2011; Peirson-Smith 2013), few have explored cosplayers' reuse of material items. However, material relationships are an intrinsic aspect of the practice as cosplayers assemble, perform, photograph, and narrate costume objects. A fuller understanding of cosplay and the practice's complex relationship with globalized production and consumption processes requires an understanding of cosplayers' relationships with material things, including their reuse activities.

As many anthropological studies of material culture have argued, the reuse of objects is deeply embedded within social relationships and processes; social contexts shape reuse practices, and reuse practices conversely shape social relationships (Miller 1987; Hansen 1995; Gregson and Crewe 2003; DeSilvey 2006). Drawing on Hegel's notion of 'objectification', Miller (1987) argues that objects and persons are constantly engaged in co-creative processes whereby objects and persons are defined in relation to one another. Miller (1998) applies this concept of objectification to consumption practices, including shopping. However, recycling, reusing, and gifting can also be associated with the remaking of self and social relations (Gregson and Crewe 2003; Norris 2004).

The reuse of material objects is an intrinsic feature of the communities and the practice of cosplay within Australia. The reuse and transformation of materials is a key aesthetic value of the practice and as such creates relationships between practitioners and shapes the activities and trajectories of individual practitioners. Reuse activities are also means by which social relationships between practitioners can be created or strengthened as cosplayers share knowledge and gift materials for reuse. Practitioners learn how to be cosplayers by learning to see objects as potential cosplay materials. The development of reuse skills can mark a practitioner's transition from novices to expert. In an inversion of professional costuming practices, the successful use of junk materials in costume construction rather than expensive craft materials is considered a hallmark of cosplay mastery.

Seeing the cosplay potential in objects: a form of 'skilled vision'

> Never underestimate anything as being able to be used in a costume. Eventually you'll get to the point where you'll view anything and everything as costume parts.
> (Cassandra, ACG Panel, author's transcript 2012)

As anthropologists have argued, the meanings and classifications attributed to objects and spaces are not fixed; rather an object can undergo numerous classification changes throughout the course of its 'social life', becoming a commodity, a possession, a ritual object, an heirloom, junk, waste, or rubbish (Appadurai 1986; Miller 1987; Hansen 1995). While classifications of 'junk' or 'rubbish' are often applied to objects deemed to be nearing the end of their social lives, these junk objects can be reinterpreted and reborn through reuse and upcycling processes (Hansen 1995; Gregson and Crewe 2003; Edensor 2005). For example, in his study of abandoned industrial sites, Edensor (2005) argues that the aesthetic value of material things can be reinterpreted. Items and spaces classified as waste, rubbish, or junk can be reinterpreted as aesthetically beautiful, as sites of history, as uncanny or surreal meeting places between past and present (Edensor 2005). Gregson and Crewe (2003, p. 116), in their study of second-hand practices in the United Kingdom, argue that "rubbish" items can be transformed and "revalorized," physically and symbolically, into "useful" objects through sorting, cleaning, and commoditization processes. Hansen (1995) identifies similar processes at work in her study of second hand clothing markets in Lusaka, as cast off clothing items donated from Western countries are remade and revalued as local markers of taste, fashion, and style.

Crocker (2012, p. 16), echoing contemporary anthropological perspectives on material culture and consumption, argues that our consumption practices are influenced by, and inherited from, others; our uses of objects are situated within the practices and routines of daily life. Interpretations or framings of material objects are not formed by individuals in isolation, but are instead community-specific and are taught and learned. The ability to discern cosplay potential in apparently mundane objects can be considered a form of community-specific "skilled vision" (Grasseni 2007). Drawing on Lave and Wenger (1991), Grasseni (2007) argues that learning to see is a social activity, and one which is tied to the formation of a person's identity as a member of a community of practice. The training of skilled vision can take place in highly formalized learning environments such as lectures and classes as well as in the everyday lives and experiences of community members (Grasseni 2007). In cosplay, becoming a skilled practitioner, a cosplayer, requires the acquisition of 'cosplay vision', the ability to see cosplay potential in unlikely objects.

The teaching and learning of skilled 'cosplay vision' takes place in several contexts: some didactic, some highly experiential. Among the most formal learning contexts are cosplay panel presentations. Panels are events at Australian popular culture conventions in which a presenter (or multiple presenters) delivers a lecture to an audience on a particular topic. A 'question and answer' session where the presenter responds to questions from audience members is also typically included as a component of the presentation. Panel-style events are included as part of nearly all Australian conventions. Panels are important teaching and learning contexts as they provide a meeting point for novices, experts, and outsiders, a context in which practitioners can reflect upon and define the nature of cosplay.

Beyond the helpful hints provided by experienced cosplayers in panels, much of the teaching and learning about reusing materials takes place in the backstage contexts of cosplay assembly. Cosplayers regularly collaborate in the construction of cosplay costumes and props. During these collaborations, cosplayers often support or teach each other how to see and reuse items as a cosplayer. They visit recycled materials and opportunity stores together and help each other transform mundane objects through exchanging transformation skills. Sometimes these collaborative activities reflect an apprenticeship model of learning (Coy 1989; Singleton 1998) as experienced cosplayers teach skills to newcomers. Cosplayer

Figure 4.1 Packing box as craft material (2015)

Photo: Claire Langsford.

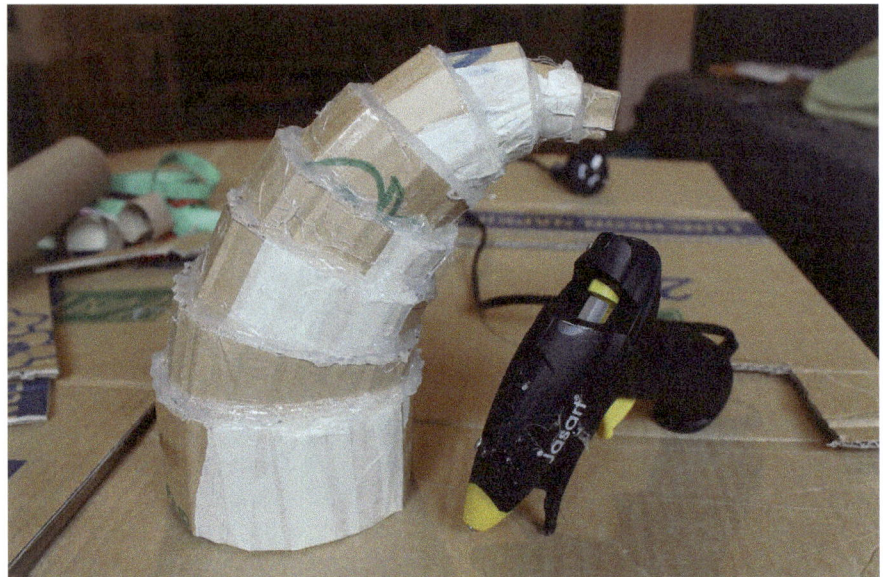

Figure 4.2 In transformation (2015)
Photo: Claire Langsford.

Figure 4.3 Painted horns (2015)
Photo: Claire Langsford.

Renée, for example, described in interview how she was learning corsetry skills from a more experienced friend as they worked with the materials together and drank coffee (Renée, author's interview 2012).

The following sections of this chapter will address the re-creation of cosplay 'skilled vision' in numerous teaching and learning contexts. It will explore how

Figure 4.4 Devil's horns (2015)

Photo: Claire Langsford.

cosplayers convey particular attitudes towards the reuse of 'junk' objects, and develop the community-specific skills which enable practitioners to transform rubbish into costumes, and to transform themselves from novices into cosplayers. Included in this chapter are excerpts of transcripts from the author's video recordings of cosplay instructional panels and competitions held at popular culture convention events, as well as in-depth structured interviews with participants.

Thrift play: cosplay discourses of reuse as thrift and creativity

To understand the role of reuse in cosplay practice in Australia it is necessary to first explore the meanings that cosplayers attribute to reuse and recycling. As in other cultural contexts, the meanings of reuse in cosplay are not fixed; instead, reuse practices are interpreted in several different ways by cosplayers. Cosplayers in Australia talk a lot about 'reuse' and 'recycling'. The concepts are discussed in panels, on blogs and online forums, in everyday conversations, and even on stage in competition performances. These discussions work to 'frame' (Goffman 1981; Giddens 1984) reuse practices and shape community-wide interpretations. Among Australian cosplayers, reuse is commonly associated with two key discourses: ideas of thrift and ideas of creativity.

C: Do you ever repurpose bits of your costumes and props?

R: Yep. Reuse, recycle; get stuff wherever you can. If you can get it for cheap then get it. That's the other thing; cosplay is bloody expensive! People spend so much money on this (Renée, author's interview 2012).

For many cosplayers, the reuse of objects and materials is framed as a necessity. The cost of recreating costumes using the materials and techniques used by professionals can be prohibitive for amateurs. Thrift plays a role in many consumption practices (Miller 1998; Gregson and Crewe 2003). It has been argued that it is the idea of thriftiness, and the importance of being 'seen' to be thrifty, which is more important to participants than any actual financial savings (Miller 1998; Gregson and Crewe 2003). In different cultural contexts, thrift carries connotations of the moral value of restraint, or even of political resistance (Miller 1998; Gregson and Crewe 2003). For these reasons, thrift can also be the subject of discursive as well as consumption practices.

Thriftiness is publicly performed and discussed at cosplay events. Cost-cutting measures and thrifty improvisations were regularly the subject of conventional panels such as "Cosplay for Poor Students and Rich Fortunates" (Sydney Manga and Anime Show 2010). In this panel, the presenters, self-described 'poor students', explained how they had created their elaborate *Final Fantasy X* costumes from hard rubbish materials and clothing sourced from opportunity shops. Among the attendees at this panel was a cosplayer who had made an elaborate Victorian-style dress entirely from newspaper and sticky-tape. Panelists and attendees modeled thrifty cosplay for newcomers and outsiders to the practice.

The idea that cosplay without reuse can be dangerously costly is also discussed onstage in competition performances. In an onstage interview with the host during the Madman National Cosplay Championship, competitor Wirru revealed that he had spent over two thousand dollars on his handmade fiberglass armor and claimed that he was now broke. Another Australian cosplayer, devoted to creating leather-based cosplays, announced in another onstage competition interview that he was retiring as his cosplay activities were affecting his ability to make his mortgage repayments.

Another practical motivation for reuse is the issue of space, namely cosplayers' limited capacity to store materials, tools, and finished costumes in their homes. The challenges of storage were made apparent to me on a visit to the home of cosplayer Daniel. In one room of his flat, cardboard boxes leaned against the wall, fabric and stuffing were strewn over the carpet, and accessories from current and previous costumes littered his desk, chair, and windowsill. Daniel told me that his fiancée was becoming increasingly annoyed by the spread of his costumes and materials across the flat. Together the couple had decided that Daniel could not create more costumes without discarding previous ones.

To preserve space, cosplayers may even repurpose materials from prize-winning and famous costumes. In an interview, experienced cosplayer Jenita described having to break up the costume she had used to compete in the 2010 Madman National Cosplay Championship, a life-sized recreation of a giant robot from Studio Ghibli's (1986) animated feature, *Laputa: Castle in the Sky* (Miyazaki 1986). Despite her strong personal attachment to her work, Jenita decided that she had no space to store the costume in its finished form and that the materials were too expensive to be used on only one costume. As a compromise to sentiment and memory, she maintained the head of the robot, with its recycled camera lens eyes, in its completed state as a keepsake. For Jenita, thrift was a powerful motivating factor in shaping her activities as a cosplayer.

However, discourses of thrift are not the only meanings attached to cosplay reuse practices. As cross-cultural ethnographic studies have highlighted, recycling practices can be positioned within discourses that have little to do with thrift or environmentalism (Hansen 1995; Saunders 2000). Reuse can be positioned within discourses as diverse as war-memory (Saunders 2000), class, taste, and style (Hansen 1995). In panels, competition performances and casual conversations cosplayers also celebrate reuse as a form of skillful creativity. The use of junk or mundane objects in costume creation is viewed as heroic improvisation:

> S: [. . .] however, if you neither have the skills, nor the money to create these [audience laugh] you can prepare for the wonders of hard rubbish collection! So, for instance, our backpack is made from four steering wheels for ah . . . video games. So, they're chopped together and the four handles are four parts there.
>
> C: If that makes any sense? And that there you've got the black white water tube from dishwashers and stuff.
>
> (Steve and Christi an, Steampunk Panel, author's transcript 2011)

In the above transcript excerpt, experienced cosplayers and Steampunk enthusiasts, Steve and Christian, narrate the assembly of a prop to provide an example of creative reuse for their audience. In their narrative, the panelists reveal that an elaborate prop – a jetpack backpack – was not made from metal and leather as it appeared to be, but was, instead, produced through the reuse of 'rubbish items'. Their audience clapped and cheered at these revelations, providing further positive feedback to the creators. In cosplay panel presentations, the reuse of junk objects is often discussed in a celebratory tone. Cosplayers who can create walking trees from papier-mâché and piping and ball gowns from second-hand bridal dresses are regularly praised for their ingenuity and skill in performing magical transformations.

In these contexts, reuse is associated with creative transformation. Cosplayers in Australia value the skill in transforming one material thing into another. At a cosplay panel in Perth, panelist Wirru recounted to the audience how he created his "Final Fantasy 13 Noctis engine sword thing" – a large and highly detailed prop weapon. Gesturing with his hand, he drew the audience's attention to the projected PowerPoint slide behind him, the slide covered with digital photographs of objects in various states of transformation – half-painted, cut-up, and sanded.

> The two toilet pipes I'm talking about! The big pipe and the small pipe – they're for one engine and there's the Lipton Iced bottle and there's a sink strainer there. Then these bits were all made out of craft foam. That was DAS – air dry clay. And just a mixture of everything in that – you get the sword.
>
> (Wirru, Wai-Con Prop Panel, author transcript 2010)

Observing the photographs displayed on the slides in conjunction with the panelist's narration, the audience perceived how the tea bottle, the plumbing pipes, and the sink strainer were incorporated into the prop. Unrecognizable in photographs of the finished props, these everyday objects had been brought back into the visual awareness of the audience by the panelist's highlighting narrative. The audience was then able to view the prop both as a whole, finished article and also as a constructed object created from unusual materials.

The most dramatic transformations seem to garner the most applause and positive comments from audience members – the more lowly, rubbishy, or junk-like the original materials, the more amazing the transformation. For the transformation to be successful, the original nature of the materials must not be visible to the eye. The original nature of the materials and the cosplayer's skill in transformation is instead performed through narratives: revelations to friends in casual conversations, demonstrations of construction methods in panels and blogs, and formal competition interview narratives, which will be explored in the final section of this chapter. These stories share similarities with "bargain boasting" narratives identified by Gregson and Crewe (2003, p. 102) in which consumers would display their outfits before verbally revealing the shockingly low price at which the items were purchased. In both bargain boasting and transformation narratives, the visual skills of the narrator are emphasized. The narrator is recounting their ability to assess the properties of an object and its component materials, their knowledge of the market and comparable objects. It is in the performance, the telling, that these creative skills become visible.

The skills involved in collecting and assembling materials

In cosplay, reuse activities are mostly conducted in the offstage contexts of the assembly or construction of the costume. For cosplayers, creative or thrifty reuse requires community-specific knowledge and skills. To put together a costume – from fabric, plastic, metal, or paper – a cosplayer must know where to source materials, how to assess those materials for "cosplay potential," and how to transform materials or objects into functional costume components. Knowing how to shop, scavenge, and borrow are key competences involved in costume assembly. Studies of consumption practices have identified shopping and other consumption activities as requiring different skills and aims for different fields – be it bargain hunting, shopping to display cultural or economic status, or to demonstrate appreciation of subcultural authenticity (Clarke 1998; Jackson et al. 2005). To be a cosplayer and to claim membership in a community of practice, a newcomer must learn how to shop like a cosplayer.

In their study of second-hand cultures in the UK, Gregson and Crewe (2003, p. 140) argue that "geographical knowledge," knowing where to shop, is an important element of strategic and successful second-hand shopping. Local cosplay geographical knowledge is centered upon identifying the best places to find objects suitable for reuse. This knowledge is shared between cosplayers in panels, online forums, and casual conversations.

A regular topic of conversation in panels and forums is the identification of 'good' charity or opportunity shops, where panelists provide personal recommendations to their audience. Particular second-hand shops are identified as tending to stock particular items suitable for reuse in cosplay, stocking quality items, or having cheap prices. In a panel lecture, Ben, an experienced cosplayer, promotes his favorite opportunity shop to his audience:

> B: Vinnies or op-shops are one of the best places I think to source things, little things sometimes and even for guys if you go into op-shops and you think, oh it's mostly women's clothing, there is particularly one op-shop at Glengowrie called Men's For You and it's an op-shop entirely of men's clothing. They know me by name now!
>
> (Ben, ACG Panel, author's transcript 2011)

Panelists also proselytize the opportunities provided by local council hard waste collections:

> A lot of us actually, when it's hard rubbish day, we tend to go scouring up and down the street, looking around and seeing what people are throwing out and

going, "How can I use that?" You never know what you'll find on the side of the road, depending on the area that you live in.

(Ben, ACG Panel, author's transcript 2011)

The home is also promoted as another place from which to source cosplay materials:

It's best to use what you've got around the place as well. This was made from using the bottoms of plastic shopping bags and just gluing them together in layers and then over the top of that we went over it with electrical wire as well and electrical tape.

(Liz, ACG Panel, author's transcript 2011)

However, knowing where to look for reusable materials is only the first step. The found materials must then be evaluated for their cosplay potential. Gregson and Crewe (2003, p. 140) identify a further skill for second-hand shopping which they term "the knack" or "the eye." "The eye" describes a shopper's ability to assess and take inspiration from the physical properties of an object itself.

"The eye" in cosplay is often associated with a cosplayer's personal relationship with materials, a cosplayer's ability to select appropriate materials, work with them by hand, and recreate details. Developing a personal relationship with materials is considered a hallmark of good cosplay within the community. In panel presentations, 'amateur' cosplay construction is often set up in dichotomy with 'professional' mass-produced, Halloween-style costumes.

Be careful of online cosplay shops as well because sometimes they take shortcuts when making their costumes to save on costs and manufacturing time, so they're not always accurate.

(Liz, ACG Panel, author's transcript 2011)

Panelists describe amateur construction as detailed, careful, and improvisatory. Mass-produced costumes are deemed to be inaccurate and impersonal, produced without a personal relationship with the tools and materials.

The ability to accurately assess the physical properties of a material is most often acquired through long-term practical experience (Gregson and Crewe 2003; Keller 2001). In panels, cosplayers often recommend newcomers to engage in experimentation with new materials before beginning costume projects. In his panel, Darren describes how he came to learn the silicon molding techniques he used for his "Cyberman" costume from the television series *Doctor Who*.

I've never really had any professional help, TAFE courses or anything to learn how to do this material. Most of this I've just grabbed the stuff and used it and worked out how to use it on there [laughs]. Before I started making this Cyberman I'd never made a large silicon mold but by the end of it I'd made sixteen molds. So you just keep getting better and better and all the mistakes you make on the first one you hopefully don't make them on the next one! [laughs]

(Darren, ACG Panel, author's transcript, 2012)

However, cosplay panels also attempt to teach newcomers to cosplay what to 'look for' when choosing materials. Using finished costumes or photographs of costumes as models, panelists 'highlight' (Goodwin 1994) features that may not be noticed or appear relevant to an untrained eye:

> You need to look at the grain of the fabric. Some, like a fabric called drill, the grain is on an angle, [holds up dustcoat] and this one here has a very definite grain. So if you have a really good look at close up shots, you get to see all these details and they can make the difference between a really, really good costume and one that looks kind of average.
>
> (Liz, ACG Panel Transcript 2011)

Here, in this example of panel performance, the presenter was highlighting: inviting the audience to view a garment, a costume dustcoat jacket, in a very particular manner. She held up the garment to illustrate with her hands the way that the fabric has been woven and also to demonstrate that the weave of the material matched the appearance of the videogame character's costume shown in the PowerPoint slide behind her.

In this way, through verbalized performance, modeling, and highlighting, panelists instruct newcomers in a cosplay way of seeing. Through sharing geographical knowledge regarding the appropriate sources of reusable materials, and teaching newcomers how to appraise potential costume objects, cosplayers school participants in a culture of reuse.

Sharing knowledge and exchanging junk: creating relationships and reinforcing reuse practices

Another means by which reuse activities are integrated into the practice is through cosplayers sharing, exchanging, and gifting costumes or reusable materials. Anthropological material culture studies have long identified gifting and exchange as means by which relationships and community values can be maintained and strengthened (Clarke 1998; Miller 1998). The sharing of objects strengthens relationships between cosplayers but also normalizes reuse and reinforces its importance within cosplay.

In certain contexts cosplayers may gift or lend entire costumes or completed costume parts. Throughout my fieldwork period, I regularly observed cosplayers sharing expensive wigs or loaning props, extending the use of the costume beyond the original purchaser or creator. A commissioned costume I purchased online was eventually used by two other cosplayers at events. The costume was a recreation of the protagonist character "Ash Ketchum" from the anime series *Pokémon*. It consisted of a jacket, wig, cap, and gloves. I lent these items on two separate occasions to cosplayers who needed an outfit for an upcoming event. The act of sharing costumes can be positioned within the broader community value of collaboration. The mutual exchange of labor, materials, and skills is valued by cosplayers and can also be seen elsewhere in the practice of the exchange of labor between cosplay models and photographers and the role of "Cosplay Medics," practitioners who attend events to help repair damaged costumes.

Cosplayers would also gift or exchange reusable materials. As a favor for sewing up a jacket, cosplayer Jane provided me with a large bag of industrial leather offcuts. In this example, the exchange worked on a personal level between two cosplayers. However, gifting or exchanging could involve large numbers of cosplayers. Experienced cosplayer Renée decided that she had too much fabric stored and decided to offer it online to other cosplayers. She created an event on her own Facebook page and advertised the fabric to other cosplaying friends.

Reused materials could become the subject of narratives connecting multiple consumers and practitioners. Anthropologist Shankar (2006) has argued that narratives told about an object and its consumption actually extend the relationship between an object and its consumer(s) to a broader network. In one case I observed during fieldwork Jane was gifted some golden curtain fabric from her sister. Her sister had purchased the fabric but decided that she could neither use it nor store it and gave it to Jane for costume making. Jane in turn gave the fabric to cosplayer Daniel, who needed golden fabric for a coat. Jane told her sister that Daniel was using the fabric for a coat. Jane's sister became interested in Daniel's construction processes and requested that he post pictures of the finished garment. In this way, the golden fabric and its transformation became a means of connecting multiple consumers and practitioners.

Spectacular transformations: the valorization of reuse in performance narratives

While reuse practices and skilled cosplay vision are predominantly taught and learned in 'backstage' contexts and costume assembly processes, the importance of reuse is also proselytized onstage in cosplay competitions. In many forms of cosplay competition, the hosts or judges of the competition will interview competitors onstage about the costumes they are wearing. Through the use of questioning, the host or judge attempts to elicit the story of the costume and how and why it was created. These interviews are considered part of the cosplay performance. They serve to entertain and educate the audience and provide insights into the creation of the costume. In these narratives, competitors tell stories of transformation, narrating how they transformed various materials into the finished costume visible onstage.

In their ethnographic account of second-hand markets, including charity shops, Gregson and Crewe (2003, p. 116) use the term "revalorization" to describe how objects previously classified as "junk" or "rubbish" are re-ascribed value as useful, personal, and meaningful objects. "Revalorization" is a particularly apt description of the re-classification processes performed during cosplay competition interviews, as cosplayers narrate the transformation of junk materials into costume objects in the form of heroic tales. In telling these tales, competitors also transform themselves into skilled cosplay masters who can reuse, reassemble, and repurpose unlikely objects into spectacular costumes.

Okabe (2012) in his ethnographic study of Japanese cosplayers argues that the sharing of 'heroic' tales of particular cosplayers' exploits is an important form of knowledge transmission among practitioners. These tales provide inspiration and a model of behavior for other practitioners (Okabe 2012). Heroic cosplay tales are also evident in Australian cosplay communities, particularly in competition interview narratives where competitors are asked to describe the construction processes for their costumes and props. Like the "bargain boasting" narratives identified by Gregson and Crewe (2003), cosplay heroic narratives hinge on the juxtaposition between the spectacular visual appearance of the costume and the mundane materials used. The cosplayer's ability to transform something mundane into something spectacular is framed here as a heroic feat.

To demonstrate how heroic narratives of transformation are constructed, this section of the chapter will focus on the case of a particular interview performed during the Madman National Cosplay Championship in Brisbane in 2010. As an audience member, I was amazed by a strikingly tall female cosplayer who was dressed as the villainous character "Cain Nightroad" from the Gonzo Studios 2005 anime series *Trinity Blood* (Hirata 2005). Arrayed in bronze and white, carrying a massive prop mace which was taller than head height, she had six feathered wings attached to her back which was further decorated by a bronze, spiked wheel which acted somewhat like a halo or aureole in religious iconography.

Host: Tell us about the wheel on your back.
A: The wheel on my back?
Host: It's not a wheel I know. I don't know what it is, it's just awesome.
A: Ah, my halo of spiky awesomeness?
Host: Oh, that's the technical name? That's awesome! Okay, your halo of spiky awesomeness. What are we dealing with there?
A: It's actually made out of Tooheys poster board. Tooheys beer, good stuff (Madman National Championship 2010 Video, author's transcript).

The audience and host laughed at this revelation. The original material, the Toohey's poster board, has been so transformed by the cosplayer that it is completely invisible to the audience, even with telephoto lenses. It is only through the verbal description of the costume that the audience is able to simultaneously picture two different visions of the object – as halo of spiky awesomeness and as beer advertising cardboard.

Host: Now, I checked the notes on this [points to decoration on back of costume] that's a real skull! This is awesome! Where did you pick up the skull?

A: Glen Innes [a NSW country town]. It's a small county town about ten
           hours train ride from Sydney.
Host: You didn't kill it, did you?
A: Course I did!
Host: [pointing at lance] With that?
A: No with my bare hands and teeth. It's the only way to go! [crowd laughs].
Host: Then I killed an animal. Then I drank some beer. That's awesome! This is
           so very, very cool.
                    (Madman National Championship 2010 Video, author's transcript)

In recounting the incorporation of the sheep's skull into her costume, the cosplayer exaggerates the process, weaving in a story of violent slaughter for comic effect. Humorous exaggerations and fictionalizations of transformation processes were often used by cosplayers in competition interview narratives to entertain audiences but also to magnify the aura of magic surrounding object transformation. Competition interview narratives therefore serve as opportunities for cosplayers to aggrandize their own skills but also function to elevate reuse activities to the position of mythic feats. These narratives and stories of transformation are later distributed throughout communities through video recordings and recounts on social networking sites, further conveying the notion of reuse as skilled creativity.

Conclusion: What can cosplayers teach us about communities and reuse?

Community-specific reuse practices can provide compelling counter narratives to Western mainstream discourses surrounding consumption. For cosplayers, the 'junk' objects left out for hard rubbish collection occupy a rather different status. They are not objects at the end of their lives, but instead represent the potential creation of new objects, potential relationships between other practitioners, or a means of gaining increased social standing in a very specific community. Cosplayers' views on reuse also set them apart from some other reusers who view reuse as primarily an environmental or political act. For cosplayers, reuse is about thrift as creativity, magical transformations, and a deep, personal engagement with materials in craftwork.

The case study of cosplayers within Australia provides an opportunity to explore a community of practice in which reuse is framed as necessity and celebrated as creative improvisation. This study echoes the findings of previous anthropological material culture studies (Hansen 1995; Saunders 2000) in showing that discourses outside of environmentalism or other ethical concerns can also underpin reuse practices. Policy-makers should consider the broad scope of possible motivations and meanings attached to reuse, even if the ultimate goal of the study or program is to promote reuse for environmental or ethical ends. By conducting investigations into existing community attitudes towards reuse they may uncover existing values and practices that may create a barrier towards reuse or alternatively, as in the case of cosplay, promote and support reuse.

Transforming producers' and consumers' attitudes towards reuse is a challenging task for policymakers, educators, and industry regulators. Promoting different forms of reuse, be it second hand shopping to fashion enthusiasts or the reuse of waste materials to building companies, is particularly difficult as attitudes towards reuse are significantly shaped by a range of structural factors. Reuse activities are always positioned within social practices, situated within a web of meanings and relationships.

This study of reuse in cosplay communities also demonstrates how reuse practices are sustained when discourses and actions surrounding reuse are embedded in every aspect of the practice. Reuse is a key aspect of cosplay community participation. Cosplayers must learn to value reuse activities as creative feats, must know where to source appropriate materials, how to appraise these objects for their cosplay potential, and how to transform them into new costume objects. Reuse cannot be an adjunct to practice but must be embedded into diverse aspects of the practice community, including its social fabric. In cosplay communities of practice, practitioners are motivated towards reuse because it is framed as a prestigious, skilled activity. Skilled reusers are celebrated as community leaders who model their skills publicly and impart their knowledge to community newcomers.

The integration of reuse into a community of practice begins with the way that reuse values and practices are taught. Throughout this chapter I have demonstrated that learning processes associated with reuse are not sidelined from everyday cosplay activities but are instead integrated throughout the main aspects of cosplay practice. Skills and knowledge are transmitted formally in panels; through sharing and gifting reusable materials cosplayers strengthen relationships and the normalization of reuse; in competition performances reuse transformations are celebrated and rendered magical.

The highly integrated nature of reuse activities in this case study suggests that those attempting to encourage reuse activities in other practices and communities need to adopt a more holistic approach to promoting and normalizing reuse, as challenging as this may be. For those working towards promoting reuse in much larger fields than cosplay – in other practices, workplaces, schools, communities, even at the level of government and industry – this study of cosplay demonstrates that reuse cannot be an afterthought. Attitudes and activities must be transformed from the ground up, and a commitment to reuse must be integrated into multiple elements of a practice.

References

Appadurai, A. (1986) *The Social Life of Things: Commodities in Cultural Perspective*, Cambridge; New York: Cambridge University Press.

Booth, P. J. (2012) "Mashup as Temporal Amalgam: Time, Taste, and Textuality", *Transformative Works and Cultures*, p. 9.

Clarke, A. (1998) "Window Shopping at Home: Classifieds, Catalogues and New Consumer Skills", in D. Miller (ed) *Material Cultures: Why Some Things Matter*, Chicago: University of Chicago Press.

Coy, M. W. (1989) *Apprenticeship from Theory to Method and Back Again*, Albany: State University of New York Press.

Crocker, R. (2012) "'Somebody else's problem': Consumer Culture, Waste and Behaviour Change – The Case of Walking", in S. Lehmann and R. Crocker (eds) *Designing for Zero Waste: Consumption, Technologies and the Built Environment*, London; New York: Earthscan.

De Certeau, M. (1984) *The Practice of Everyday Life*, Berkley; Los Angeles; London: University of California Press.

DeSilvey, C (2006) "Observed Decay: Telling Stories with Mutable Things", *Journal of Material Culture*, vol 11, no 1, pp. 318–338.

Edensor, T. (2005) "Waste Matter – The Debris of Industrial Ruins and the Disordering of the Material World", *Journal of Material Culture*, vol 10, no 3, pp. 311–332.

Giddens, A. (1984) *The Constitution of Society: Outline of the Theory of Structuration*, Cambridge: Polity Press.

Goffman, E. (1981) *Forms of Talk*, Philadelphia: University of Pennsylvania Press.

Goodwin, C. (1994) "Professional Vision", *American Anthropologist*, vol 96, no 3, pp. 606–633.

Grasseni, C. (2007) *Skilled Visions: Between Apprenticeship and Standards*, Oxford; New York: Berghahn Books.

Gregson, N. and Crewe, L (2003) *Second-Hand Cultures*, Oxford; New York: Berg.

Hansen, K. T. (1995) "Transnational Biographies and Local Meanings: Used Clothing Practices in Lusaka", *Journal of African Studies*, vol 21, no 1, pp. 131–145.

Hirata, T. (2005) *Trinity Blood* (anime series for TV), Tokyo: Gonzo Studios.

Jackson, P., Rowlands, M. and Miller, D. (2005) *Shopping, Place and Identity*, London: Routledge.

Jenkins, H. (2013) *Textual Poachers: Television Fans and Participatory Culture*, New York: Routledge.

Keller, C. M. (2001) "Thought and Production: Insights of the Practitioner", in M. B. Schiffer (ed) *Anthropological Perspectives on Technology*, Oxford; New York: Oxford University Press.

Lamerichs, N. (2011) "Stranger Than Fiction: Fan Identity in Cosplay", *Transformative Works and Cultures*, p. 7.

Lave and Wenger (1991) *Situated Learning: Legitimate Peripheral Participation*, Cambridge; New York: Cambridge University Press.

Lunning, F. (2011) "Cosplay, Drag and the Performance of Abjection", in T. Perper and M. Cornog (eds) *Mangatopia: Essays on Manga and Anime in the Modern World*, Santa Barbara; Denver; Oxford: Libraries Unlimited.

Miller, D. (1987) *Material Culture and Mass Consumption*, New York; Oxford: Basil Blackwell.

Miller, D. (1998) *A Theory of Shopping*, Ithaca, NY: Cornell University Press.

Miyazaki, H. (1986) *Laputa: Castle in the Sky* (anime video) Tokyo: Ghibli Studios.

Norris, L. (2004) "Shedding Skins: The Materiality of Divestment in India", *Journal of Material Culture*, vol 9, no 1, pp. 59–71.

Okabe, D. (2012) "Cosplay, Learning and Cultural Practice", in M. Ito, D. Okabe, D and I. Tsuji (eds) *Fandom Unbound: Otaku Culture in a Connected World*, New Haven and London: Yale University Press.

O'Reilly, K. (2005) *Ethnographic Methods*, London; New York: Routledge.

Peirson-Smith, A. (2013) "Fashioning the Fantastical Self: An Examination of the Cosplay Dress-up Phenomenon in Southeast Asia", *Fashion Theory*, vol 17, no 1, pp. 77–112.

Saunders, N. J. (2000) "Bodies of Metal, Shells of Memory: 'Trench Art', and the Great War Re-cycled", *Journal of Material Culture*, vol 5, no 1, pp. 43–67.

Shankar, S. (2006) "Metaconsumptive Practices and the Circulation of Objectifications", *Journal of Material Culture*, vol 11, no 3, pp. 293–317.

Singleton, J. (1998) *Learning in Likely Places: Varieties of Apprenticeship in Japan*, Cambridge; New York: Cambridge University Press.

Winge, T. (2006) "Costuming the Imagination: Origins of Anime and Manga Cosplay", *Mechademia*, vol 1, pp. 65–76.

Part 2

Strategies and landscapes of reuse

5 Renew(ing) Newcastle and complicating capitalism

Contributory economies, artisanal production, and the DIY occupation of disused commercial space

Cathy Smith

Introduction: the new DIY production ethic for the twenty-first century city

Capitalism and consumerism are often positioned as persuasive and intractable global forces singularly inflecting all aspects of contemporary life (Gibson-Graham 2006), including the built environment (Harvey 2010; Augé 1995). Yet the French philosopher Bernard Stiegler (2011; Stiegler and Fayner 2013) has posited an evolving relation between consumerism and urbanization in the twenty-first century city, which complicates the sense that capitalism has a uni-vocal character. The "economy of contribution" (Stiegler and Fayner 2013) is a socially-orientated approach to production and consumption intertwined with a Do It Yourself (DIY) production ethic. In the popular media, Stiegler's "new industrial model" (Lemmens 2011, p. 39) is variously referred to as the "sharing economy" (Marshall 2014, p. 22), "indie capitalism" (Bianchi and Maffei 2012, p. 6), and "hipster economics" (Kendzior 2014) because its "bottom-up" (Stiegler and Fayner 2013, p. 2) production methodology is linked to individual artisans rather than large-scale manufacturers.

If this argument is to be believed, the contributory model is the inevitable suc-cessor to traditional consumerism because it reinstates mutually beneficial inter-actions between consumers and production systems. Stiegler (2010a, p. 5) argues that mainstream consumerism has an inbuilt chronological obsolescence because it promotes "short-term," detached, and unsustainable relations between people and their disposable artifacts. By contrast, the economy of contribution promotes a sense of social and cultural interconnectedness because consumers have some degree of direct input into the design and/or production of the artifacts and spaces they consume (Stiegler 2011, p. 41, p. 126; Stiegler and Fayner 2013, p. 2). Stie-gler (2010b) also suggests that the contributory model is intractable from forms of regional and urban development that prioritize "social inclusion and authentic-ity" through localized production and consumption. While no specific examples of urban formations are cited, reference is made to technologies and production techniques which support project self-initiation and self-production, includ-ing "fab labs" (free access fabrication laboratories), online knowledge-sharing communities, and "open source" design platforms (Lemmens 2011, pp. 40–41; Stiegler 2011, p. 126; Stiegler and Fayner 2013, p. 2). In a specific example of the contributory model, Stiegler (Stiegler and Fayner 2013, p. 3) and his collaborators from the School of Decorative Arts in Paris developed a theoretical business model for a fashion company formed exclusively of consumers involved in all aspects of design and production.

With limited examples of the contributory economy to date, there is little indication of what differentiates contributory practices and urbanism from their mainstream counterparts. Thus, to expand upon Stiegler's (Stiegler 2010b; 2011; Stiegler and Fayner 2013) theorizations, reference will be made to other cultural, design, and philosophical discourses which similarly explore the interconnections between production, consumption, urbanization, and sociality. In what follows, there is a specific concentration on a particular project of DIY urbanism: the self-initiated Renew Newcastle (RN) scheme in the post-industrial inner city of Newcastle, Australia. RN is both a not-for-profit organization and approach to the problem of urban, cultural, and social decay (Renew Newcastle 2016). It provides a distinctive case for the repurposing of 'failed' capitalist spaces as part of a grassroots social agenda achieved through the temporary insertion of localized artisanal production and artistic practices within vacant commercial interiors.

RN is of particular interest to the present chapter because of its discernible resonances with key aspects of Stiegler's (Stiegler and Fayner 2013) contributory economy. With its primary social agenda to reactivate the city through artistic and cultural practices, RN's success is attributed to its coexistent online and physical existence within a very particular, post-industrial urban milieu (Westbury 2010; Renew Newcastle 2016). RN facilitates 30-day rolling participation agreements between its local creatives and willing property owners who pay an inexpensive participation fee until (or if) a commercial tenant is secured by the property owner (Creative Industries Innovation Centre 2010, p. 2; Renew Newcastle 2016). While RN projects invoke the potentiality and ongoing viability of particular commercial buildings that have lost their former business tenants, their creative occupations are only possible as a direct consequence of the failures of the commercial spaces in which they reside. Indeed, RN's artisans cannot compete with local adjacent businesses and must conform to the definition of a "creative or cultural enterprise" (Renew Newcastle 2016). As such, RN's operations cannot be positioned as a straightforward alternative to mainstream consumerism and urbanism, nor historically disconnected from the earlier cooperative mining practices of its host city – its hybridity and coterminosity with mainstream production complicating its assimilation into "capitalism's ubiquity" (Gibson-Graham 2006, p. 261).

RN's DIY urbanism will be explored through reference to other philosophical and historical accounts of artisanal production bound to urbanization. In order to invoke the territorialization, organization, and subjugation of bodies, labor, matter, and land within the capitalist apparatus, Stiegler (2010a, pp. 32–33) refers to the writings of Gilles Deleuze and his collaborator, psychoanalyst Félix Guattari. In turn, Deleuze and Guattari (2014) draw directly from the writings of their twentieth century contemporary, French economic historian Fernand Braudel, and as such, the present chapter also refers to Braudel's (1982 [1979]) positioning of localized artisanal economies as instances of pre-industrialized capitalism and urbanism. Braudel's (1982) and Deleuze and Guattari's (2014) accounts of capitalism indicate an inevitable though productive relation between artisanal production, mainstream capitalism, and urban forms; a relation that is not only detectable within RN's DIY urbanism and its particular contributory methodology, but which – through its coexistence with mainstream capitalism – complicates the notion that capitalism and consumerism have an indivisible character.

The DIY Urbanism of Renew Newcastle and its contributory production methodologies

The term "DIY urbanism" is used in cultural, urban, and architectural discourses today to describe a divergent range of community-focused, self-initiated, and self-produced temporary interventions in existing cities (Zeiger 2011, p. 2; Deslandes 2013, p. 217). The nomenclature of DIY urbanism also extends to the temporary occupation of vacant urban buildings, such as those associated with Renew Newcastle (RN) in Australia and other similar schemes in North America and Britain (Deslandes 2013, p. 218). According to architect and Wikihouse co-founder Alastair Parvin (2013), the DIY construction of the city by its own citizens is a post-Marxist, post-industrialization production methodology because responsibility for production rests with the community who initiates, makes, and occupies the resultant spaces. DIY urbanism is closely associated with the production ethic of the contemporary 'maker movement' or 'third industrial revolution' – the successor to both the first industrial and second digital information revolution (Yang and Lai 2014).

This 'revolution' is dependent on, and inseparable from, contemporary digital networks which enable consumers to directly access the requisite design and technical information required for self-production, without the direct input of professionally qualified designers and makers (Lemmens 2011, p. 39; Parvin 2013; Ratti and Claudel 2015, p. 77). For example, "fab labs" (Stiegler and Fayner 2013, p. 5) and DIY "hackerspaces" (Lindtner 2014, p. 14) enable local community members to rapid-prototype their own creations and/or modify open-source designs using the digital printers, computer-controlled laser cutters and the basic hand tools often provided in these public spaces. RN's founder, former 'Novocastrian' arts event manager and media presenter, Marcus Westbury (2013b), also posits online social media as both an enabling medium and metaphor for RN's own DIY urbanism: "Out there online there is a maker-based, design-led, web- and app-enabled DIY, small-scale creative revolution that is taking place virtually everywhere." The variety of projects associated with DIY urbanism arguably contribute to its nebulousness as both a notion and practice: project types include installation art, group performances, community gardens, guerrilla knitting, and seeding-bombing, to name a few (Figure 5.1). Regardless of their typological diversity, the projects of DIY urbanism are instantiated outside of master-planned and

Figure 5.1 The Fence Parasite (2014–2015), by Cathy Smith and Rowan Olsson: an example of a self-initiated DIY Urbanism project in Newcastle, Australia; involving the co-option of street pavement for neighborhood events using its foldable stools and table (2014)

Photo: Cathy Smith.

sanctioned financial, legal, and institutional frameworks, even when projects involve professional designers, planners, and architects (Taylor 2011, p. 47).

Due to DIY urbanism being coterminous with mainstream urbanism, there is an ongoing concern about the potential assimilation of its grassroots practices into mainstream urbanism and commerce (Zeiger 2012; Deslandes 2013, p. 218; O'Callaghan 2014, p. 2; Kendzior 2014; Finn 2014, p. 394). While the projects of DIY urbanism may not always generate a discernible economic output, the RN scheme has been linked to 'potential' (Zeiger 2012) economic benefits for both its artisans and Newcastle city. Despite this potential, there are only some tentative studies linking the RN scheme to improved local business or property valuation (Flanagan and Mitchell 2016, p. 38). Importantly, RN's stated focus and remit is neither business nor property incubation, but rather the cultural and social urban remediation of a derelict regional city (Westbury 2011; Renew Newcastle 2016). RN has, however, had a measurable impact on the social and cultural regeneration of Newcastle's formerly derelict business district (Harris 2010, p. 18; Integrated Design Commission SA 2012, p. 7). It is also difficult to consider its DIY urban formation in isolation from the flows of materials, capital, and labor of its host city.

Philosopher Manuel De Landa (2000, p. 76) speaks of historic coal-based network cities like Newcastle "as veritable transformers of matter and energy [which] needed to input flows of iron ores, limestone, water, human labor, and coal, as well as to output other flows." Using this conception, we might understand Newcastle through the flow, de-territorialization and recoding of matter in the city during its successive periods of post-colonization and post-industrialization. Located along a rich coal seam, Newcastle's industrial economy was founded upon a fractured earth after the official 1797 discovery of coal by British colonists (Andrews 2009, p. 9) – its current CBD reveals traces of mine subsidence from its many former collieries, and a filigree of rail networks servicing the export of coal and steel through the working harbor. In the Newcastle of 1999, the industrialized "matter-flow" (Deleuze and Guattari 2014, p. 451) was interrupted. The departure of the BHP steel works prompted a decade of economic decline and de-investment in the inner city, including the relocation of major retail tenancies to peripheral suburban developments (McCarthy 2010, p. 4). One might argue, following Braudel, that capitalism has a life of its own, following and choosing its targets very specifically. In Braudel's (1982, p. 445) words, capitalism "is able to choose the areas where it wants and is able to meddle, and the areas it will leave to their fate, incessantly reconstructing its own structures from these components, and therefore little by little transforming those of others." At the turn of the third millennium, Newcastle's own urban 'fate' appeared to be intertwined with the socio-economic life of its homogenous suburban shopping centers.

Nevertheless, the subsequent dereliction of the CBD prompted new contributory models of economic activity, including the self-positioned "DIY urban renewal" (Westbury 2013a) and "bottom-up approach" (Creative Industries Innovation Centre 2010) of RN. From its inception in 2008, RN's agenda was to recuperate a discernibly "dying city" (Westbury 2011) through the temporary occupation of abandoned shop fronts and commercial tenancies by budget-restricted artists and artisans financially precluded from accessing commercial properties (Westbury 2013a). Eschewing the "Central Place hierarchies" (De Landa 2000, p. 76) of a master-planned urbanism, RN's physical existence can only be temporarily mapped as it involves an unpredictable network of available space, willing landowners, and a local producer-tenant. Its inexpensive "Participation Fee" (Renew Newcastle 2011a) enables artists and makers to test ideas and artistic works (saleable or otherwise) without the usual risks associated with expensive, long-term lease commitments (Westbury 2013b). The scheme is one of simple infrastructure, architecture, and overheads; its temporary and transportable interiors are installed by their tenant-producers with little or no alteration of the existing building shell, in minimal time, and with minimal budget (Figure 5.2). Tenants typically construct and customize their artistic works inside their shop and office

Figure 5.2 Renew Newcastle's most prominent space until 2017, The Emporium, inside a prominent building formerly occupied by Australian retail giant David Jones. Launched in November 2012, The Emporium closed on 7 June 2017 due to the site's impending redevelopment. (2014)

Photo: Cathy Smith.

spaces, making labor an explicitly visible aspect of any commercial transaction that might occur within a RN space (Figure 5.3). The scheme's prime objective remains civic rather than mercantile: the social reanimation of its formerly derelict host city (Renew Newcastle 2016). In this sense, the RN focus on sociality strongly resonates with the figure of the 'amateur' or do-it-yourself of the contributory model "primarily motivated by their interests rather than by economic reasons" (Stiegler and Fayner 2013, pp. 2–3).

Stiegler, the demise of consumerism and the rise of the urban do-it-yourselfer

Stiegler (as cited in Lemmens 2011, p. 39) refers to the contributory economy as a form of "cooperative capitalism" because community members both create and consume locally-made products and services. Although Stiegler's focus is on a "*new form* of capitalism" (Stiegler in Lemmens 2011, p. 39), both the notion and its association with urbanism have discernible historical precedents. The Rochdale Equitable Pioneers Society advocated for "Consumer Cooperation"

Figure 5.3 Renew Newcastle participant, the 33 Degrees South soap factory, made and sold their cold-pressed soap in a previously vacant retail space in The Emporium. (2015)

Photo: Cathy Smith.

in nineteenth century England and its principles were used in the creation of Frank Lloyd Wright's Usonian housing scheme in North America (Reisley and Timpane 2001, pp. 10–11). The sharing of both "individual and communal risk" among all members of a cooperative was seen to produce "a mutuality of purpose" (Reisley and Timpane 2001, p. 11). For a pre-digital historical precedent for the contributory economic model, Stiegler (Stiegler and Fayner 2013, p. 2) also refers to the French international retailer Fnac (Fédération Nationale d'Achats des Cadres), initially founded in 1954 by two members of the Young Socialist movement; customers were seen as participants who directly contributed ideas and suggestions to the business and product direction. This differentiation of cooperative capitalism from other forms questions singular or unified conceptions of capitalism and consumerism which invoke "the image of a worldwide and conceptually monolithic market . . . the representation of the mass of humanity in thralldom to a singular economic structure" (Gibson-Graham 2006, p. 243).

The contributory economy also challenges traditional Marxist accounts of capitalism which suggest that the fiscal ethic of capitalism is irreconcilable with social objectives and community well-being. According to the latter, conceptualization of traditional or mainstream capitalism, all production must be disconnected from small-scale, localized markets in order to generate the unimpeded flows of materials, goods, and capital necessary for capitalism's successful expansion into global markets (Deleuze and Guattari 2008, p. 142). This process of detaching production from local milieus is evident in the following comparison of traditional artisanal and industrialized production approaches. In localized production methodologies, artisans produce entire objects for sale and are thus involved in all aspects of their production and, in many cases, retail. By contrast, industrialized workers are partial contributors to larger production systems over which they have little control or connection. Industrial production methodologies also disable the discretion of labor and its outputs by transforming artisans into indentured workers or laborers disconnected not only from consumers, but from the power to control their own laboring bodies (Stiegler 2010a, p. 33). According to Stiegler (2010a, p. 33), industrialization produces a type of bodily disconnection and "forgetting" in its workers, because automated production replaces human action by copying and then reproducing the bodily "gestures" of artisans who subsequently have little opportunity to enact these gestures themselves. To borrow from the words of Henri Lefebvre (2014, p. 33), writing on the fragmentation of labor and laboring bodies during architectural production: "Most people ignore their body and misunderstand it. Some, caught up in the division of labor, only make use of the gestures of fragmented labor, gestures that have an influence outside work and shape the body and daily life." For Stiegler (2010a), this process of disconnecting workers from their bodies, producers from consumers, is a key problem of capitalism because it invades all aspects of life, promoting a sense of fragmentation and detachment. In this sense, traditional capitalism dictates and overregulates bodily behaviors, desires, and capital along with "the flows of production, the flows of means of production, of producers and consumers" (Deleuze and Guattari 2008, p. 142).

If traditional capitalism has transformed human free action into the artificially segregated notion of 'labor', the contributory model potentially reaffirms it by restoring the operational connections between production and consumption, consumers and the artifacts, or places they self-produce (Figure 5.3). Accordingly, the consumer transforms into a new hybrid subject, the do-it-yourselfer described variously by Stiegler (Stiegler and Fayner 2013, p. 2) as the "amateur," "producer-consumer," or "contributor of tomorrow." Unlike the industrial worker, the do-it-yourselfer can be involved in all aspects of production from project initiation through to construction and use. Her vested interests in production systems are part of broader processes of self-actualization and identification (Stiegler 2010a, p. 70). As is characteristic of the many forms of DIY pursuit – mainstream DIY, countercultural DIY, and DIY urbanism, to name a few – the contributory model also involves productive intersections between what is made, purchased, and con-sumed; its attendant ethic of care replaces the ethic of carelessness characterizing consumerism's relentless and ultimately unsustainable orientation towards finan-cial gain alone (Stiegler 2011, p. 18). The emerging and positive social, cultural, and ethical outcomes of the contributory economy have led Stiegler to conclude that traditional capitalism has been superseded by cooperative capitalism and "consumerism has had its day" (Stiegler and Fayner 2013, p. 3).

Consumerism's mutation into a cooperative economic model is, for Stiegler, specifically evident in entrepreneurial artistic practices which inflect regional identity and urbanization. He suggests that "[a]rtists and creators help cities and regions create a sense of social inclusion and authenticity. Economic prosperity relies on cultural, entrepreneurial, civic, scientific, and artistic creativity" (Stie-gler 2010b). Of note is Stiegler's (2010b, pp. 1–2; 2011, pp. 9–10) differentiation between the culture industries – an extension of capitalism in their central focus on profit – and other creative practices primarily orientated towards social and cultural transformation. The key difference between the culture industries and contributory practices appears to be that the latter are "bottom-up" or consumer-led in their conception and operation (Stiegler and Fayner 2013, p. 2).

Hybridity, coexistence, and contribution in Renew Newcastle (RN)

I have previously argued that RN resists straightforward binary characterizations because of its social focus, its simultaneity with ideologically divergent forces of conventional urban planning and development, and its unique admixture of pro-duction and consumption, artisanal, and artistic practice (Smith 2015). As a further indication of its hybridity, RN's success has also been attributed to its coexistent analogue and digital existence (Westbury 2010) – a point resonant with Stiegler's (2011) own argument that contemporary urban economies cannot be separated from digital networks. While Westbury (2010) reinforces that RN's physical pres-ence in the city is essential for its host city's cultural and social remediation, it was first posited as an idea and 'social media project' through Facebook. Following initial rejections by arts funding bodies, the overwhelming community support evident in the Facebook campaign enabled RN to secure institutional support by 2009 (Westbury 2010).

RN's own inherent complexity, hybridity, and coterminous existence are par-ticularly evident in its relation with institutional funding bodies. Since 2009, RN has been supported by a range of public and private sponsors, including the NSW

State Government through its "Premier's Rural & Regional Grants Fund" (Renew Newcastle 2011b). Its partnerships have now been recognized as an exemplary business and economic model through a number of significant awards. In 2010, RN and the GPT Group – a private property portfolio manager – "won Australia's highest award for a partnership between a business and an arts organisation," awarded by the Australia Business Arts Foundation (AbaF) (Renew Newcastle 2010). The Renew Newcastle/GPT Group partnership also won the Toyota Community Partnerships award and the overall Partnership of the Year. The scheme's most prominent tenancy, the Emporium, is located in a heritage building now conjointly owned by the GPT Group and Urban Growth, the NSW State Government's property developer (NSW Government 2014). And regardless of its grassroots character, RN has emerged as an established national and international model for urban and cultural regeneration (Creative Industries Innovation Centre 2010; Westbury 2011).

The successful conjunction of grassroots, state-based, and business entities generates inevitable tensions because of the potential conflicts between property development, fiscal gain, and the independent governance expected of a public institution (Gordon 2014). On the one hand, it could (and has been) argued that RN's current funding obligations, reliance upon private land holders and temporality, expose it more readily to being subsumed by mainstream business entities, thus eroding its community basis and contributory character (Munzner and Shaw 2014, pp. 2, 16). The preconceptions of the property owners who agree to RN's temporary occupations could also potentially exclude some experimental projects which lack an easily marketable product or outcome. While the initial projects of 2008–2009 included more experimental art environments, the tenancy type has shifted alongside its expanding profile and now includes creative office-based practices and commercial, albeit artistic, retail (Figure 5.3). Without the scheme, Newcastle would currently lack the artisanal identity associated with its urban, cultural, and fiscal rebirth. A 2010 report by the local Newcastle City Council and the NSW State government recognized the "increased activity" in the CBD prompted by RN's initiatives (Harris 2010, p. 18).

RN is also cited as a key influence upon the city's emerging international identity as a city "even cooler than Seattle" (Barrett 2012). As noted by philosopher and Deleuzian scholar Craig Lundy (2012, pp. 109–110), it may be difficult to disassociate the notion of the State from the minority and vice-versa due to their coexistent processes and operations – thus "what is captured has an effect on the capturer." And while RN is specific to a particular urban milieu, its success has prompted the creation of another entity, brand, and mode of operation – that of "Renew Australia," a "National social enterprise" formed in 2012 to potentially replicate its successful regenerative model in other suitable urban sites (Renew Australia 2015). While RN can be differentiated from its sibling entity, it cannot be completely disassociated from its rhizomatic tendencies, however positive these outcomes may be; a further manifestation of RN's simultaneous resistance to, and operational reliance upon, the urban and social networks frequently associated with the global spread and success of capitalism.

On the other hand, RN's model is a place-specific response to a particular form of financial and cultural decay and is not necessarily exportable into other locales (Westbury 2015, p. 161). RN's artisanal production is also noticeably different from standardized, industrialized production methodologies synonymous with the scale and portability of traditional consumer goods, primarily because its local artisans self-manufacture their products inside their publically-accessible spaces. The scheme's funding obligations also preclude certain participants who are seen to be competitive with existing city businesses, or who do not fit the definition of an arts or cultural provider (Renew Newcastle 2011a). Importantly, RN continues to incorporate project examples without a retail component, such as the temporary installations and the exhibition of experimental content inside of the gallery areas of The Emporium: RN's most prominent interior and 'shop front' (Figure 5.4). These temporary artistic works could be described as "noncapitalist commodity production – independent commodity production by . . . collective and communal enterprises" (Gibson-Graham 2006, pp. 245–246). Accordingly, RN's social and cultural focus alongside its temporary and interventionist nature ensure its differentiation from mainstream business development (Smith 2015, p. 621).

To complicate RN's relation to traditional capitalism and urbanism, the architectural structures which appear to sustain and perpetuate its particular form of

Figure 5.4 Conditions and Speculations (2014), The Project Space, Renew Newcastle. A temporary exhibition of work by first year students of the Master of Architecture, the University of Newcastle, prompting public debate about the city of Newcastle

Photo: Cathy Smith

DIY urbanism are the empty and abandoned shells of mainstream capitalism. For Braudel (1982, p. 373), urban forms are always "parasitical formations" because they rely upon rural areas to supply materials. One could also describe the RN scheme as a productively 'parasitical' formation – "altering the system from within, contributing to our understanding of the relationship between technology, use, production, society, activism and the State" (Lindtner 2014, p. 145). RN's physical existence is deliberately interventionist and ephemeral because its long-term occupation of the city is unnecessary if the properties in which it resides become commercially viable (Creative Industries Innovation Centre 2010: Renew Newcastle 2016). Its occupations involve the overt display of production and con-sumption, process, and product, both within and alongside (rather than in com-petition with) the processes and operations of mainstream consumerism (Renew Newcastle 2016). As its name suggests, the scheme has "renewed," not replaced, Newcastle's inner-city milieu (Finney 2012, p. 72).

The coexistence of a grassroots movement with mainstream business and urbanization is not necessarily an ideological limitation or compromise per se, but instead highlights the difficulty of understanding its operations through straightforward dialectical structures. Nor should DIY urbanism's coterminous presence with mainstream planning necessarily be an argument for its incor-poration into the operational schema and agency of state-sanctioned urbanism, an argument suggested by urban theorists Kurt Iveson (2013, pp. 947, 954), Donovan Finn (2014, p. 395), and Keiken Munzner and Kate Shaw (2014). This is because, as seen in the specific case of RN, DIY urbanism may be potent 'precisely' because it is simultaneously independent and interdependent, both resistant to, yet strangely reliant upon, the processes of mainstream capitalism and urbanism that perpetuate the same urban decline it addresses. While Mun-zner and Shaw (2014, p. 17) target RN for its failure to be a "primarily economic development" strategy securing long-term tenancies for its artists, RN's tempo-rality enables a measure of experimental and incremental risk-taking often pre-cluded by large-scale financial investment and infrastructure (Westbury 2013b). Its success reinforces that alternative, socially-nuanced incarnations of consum-erism are possible alongside artistic practices lacking a discernible financial output, such as the ephemeral installations (Figure 5.4). While the "capitalist axiomatic closes and defines – in the sense of fully inhabiting – social space" (Gibson-Graham 2006, p. 89), it is not inescapable. Indeed, the simultaneity of commodified and non-commodified production within the RN scheme invokes "the indeterminate potentiality of noncapitalisms. In this space, we might iden-tify the range of economic practices that are not subsumed to capital flows" (Gibson-Graham 2006, p. 90).

Historical precedents: artisanal mining, urbanization, and the capitalist apparatus

For many urban theorists, one of the key issues of urbanization is its problematic historical role in instituting the segregations of labor associated with capitalism's subjugation of all other modes of economic and social endeavor. This is because the globalizing tendencies of capitalism and its attendant organizational super-structures (including merchants, state-based, and institutional entities) often erode the connections between localized production, urban forms, and particular con-texts or milieus (Deleuze and Guattari 2014, p. 505). The latter disconnection is exemplified in those franchised business operations and global retail chains inserted into towns without any acknowledgement of their host town's local char-acter. Similar arguments are made by anthropologist Marc Augé (1995) and urban geographer David Harvey (2010); both independently argue that architecture and urban development support the fiscal conditions and the large-scale organization of labor essential for capitalism's expansion. Braudel (1982, p. 373) also sees the large city as "the place that institutes the oldest and most revolutionary division of labour: the fields on the one hand and the activities described as urban on the other." Consistent with Marx, Deleuze, and Guattari (2014), Braudel (1982, p. 10) describes capitalism as an all-consuming entity – it is "protean, a hydra with a hundred heads."

Yet unlike Augé and Harvey, Braudel (1982, p. 373) also argues that the town or the city can resist the flows of generic capital and goods, a position further reinforced by Deleuze and Guattari in *A Thousand Plateaus* (2014). This is because certain urban forms produce their own localized milieu and identity and can therefore inhibit "the general conjunction of decoded flows" that are neces-sary for capitalism's spread and success (Deleuze and Guattari 2014, p. 505). For example, Deleuze and Guattari (2014, p. 677) refer to "alternative practices" and "urban community networks" as forms of resistance to the capitalist apparatus. As noted by geographers Katherine Gibson and Julie Graham (Gibson-Graham 2006, p. 89) these alternative practices provide an important "glimpse of what might lie outside the flows of capital." Indeed, it may be the tendency to posit capitalism as a monolithic force which itself perpetuates an ongoing injustice and "violence" (Gibson-Graham 2006, p. 90) through a failure to acknowledge and celebrate nuanced forms of urbanism that aren't synonymous with a primary focus on profit.

As seen in the case with Renew Newcastle, not all forms of urbanization are necessarily synonymous with the erosion of localized production and social ben-efit. Indeed, the complex coexistence of grassroots artisanal production alongside mainstream commerce has a long historical trajectory expressed in the complex binate relations between the town and the state, local artisanal and global produc-tion. For Deleuze and Guattari (2014, p. 506), these "polarizing forces or instru-ments of polarization" underpin all urban formations, current and historical; these are evident in the early tensions between artisanal producers, merchants, and Town or State regulatory authorities in pre-industrialized European towns. Braudel (1982, p. 62) also argues that the first urban shops were the "workshops" built by and for independent self-producing artisans and artisanal collectives resisting the regulatory oversight associated with the European artisanal markets of the time. The artisan-state production nexus associated with urbanization is also histori-cally intractable from the mining industry in both its pre- and post-industrialized forms (Braudel 1982, p. 52; De Landa 2000, p. 77; Deleuze and Guattari 2014, p. 479). Deleuzian philosopher Manuel De Landa (2000, p. 77) suggests that the two "meshworks" of heavy and artisanal or skilled industries can intersect for specific historical periods, particularly through the flows of money and capital. Prior to the industrialization of mining, miners are described as "independent artisans" (Braudel 1982, p. 52) working in small groups close to the Earth's surfaces. These earliest artisan-miners were effectively bound to the "pure productivity [. . .] of the 'subsoil'" (Deleuze and Guattari 2014, pp. 479–480). It was only during the industrialization of mining in fifteenth and sixteenth century Europe that artisan-miners became workers who exchanged their bodies and labor for an income, thus creating the discretized labor market now synonymous with mainstream capitalism (Braudel 1982, p. 52).

Following the aforementioned history of mining and artisanal practice, one might consider the RN scheme through the conceptual framework of the artisanal logic, processes, and flows of the earlier cooperative mining practices once eponymous with its host city's identity. After the discovery of coal by the British Newcastle colonists, most mining operations were conducted through state-sanctioned business entities and landowners. There was a notable exception: the worker-created colliery and associated collective, the Co-Operative Coal Mining Company formed in 1861 (Andrews 2009, p. 251). It operated until 1869 when it was purchased by a private investor, Mr. William Laidley, who had the investment capital necessary for the heavy machinery required to make the mine financially viable (Keily and Shaw 2013, p. 23). Though short-lived, this miner-owned and operated scheme was associated with other Newcastle retail cooperatives, and the "socialist ideas of co-operative community welfare" (Keily and Shaw 2013, p. 23) more broadly. Borrowing from the words of De Landa (2000, p. 77), we might posit Newcastle's economic and cultural character as inseparable from both its "large-scale, energy-intensive industry and [the] small-scale, skill-intensive industry" of RN, the latter borne of a social and cultural agenda challenging capitalism's purported indivisibility through its penultimate focus on profit.

Fluid Urban forms, reuse, and coexistence

This brief engagement with the DIY urbanism of RN, explored through Stiegler's notion of the contributory economy and Braudel's historical account of artisanal production and urbanization, indicates its rather complicated entanglement with historical processes, mainstream consumerism, and urbanization. Although RN may be a very specific example of a contributory economy, its DIY urbanism complicates any sense that capitalism is a singular entity or protean structure because it both incorporates a particular incarnation of consumerism associated with artisanal production while simultaneously refuting a focus on fiscal gain through its 'non-capitalist' projects. Moreover, its concurrence with mainstream capitalist markets, state investment, and urbanization questions Stiegler's contention that the contributory economy has subsumed mainstream consumerism. If understood as a design methodology for an enduring urban renewal, one might argue that RN presents a paradox – it exists only through the perpetual failures and abandoned architectures of capitalism. Yet an alternative economic and urban practice has been forged in those same spaces that capitalism has currently deemed without measure or value.

Newcastle's contributory practices productively reside within the urban spaces borne of the capitalistic apparatus generating novel social, economic, and urban effects without necessarily displacing the processes and products of the consumerist model. We might understand RN's particular and emerging DIY urban formation as one of several, as the intentionally temporary restitution of the material life and productivity of Newcastle's CBD through publically visible interior interventions; as the confluence of coal dust and coffee (Newcastle's 'new black'); and ultimately, as the product of matters carefully unmade and remade from its deeply fractured earths. The broader challenge for any DIY urbanism is whether it will become yet another head of capitalism's protean structure and thus cease to invoke an alternative economic model and urbanism – effecting an "absolute deterritorialisation" (Deleuze and Guattari 2014, p. 80) of RN's identity and the particular character of the city more broadly.

That said, the coterminous existence of DIY and mainstream urbanism and capitalism does not in itself predicate (nor perhaps exclude) a substitution of one system for another, nor does it support the argument that capitalism involves an inescapable singular axiomatic force expressed in the formation of the city. Rather, as argued by Gibson-Graham (2006), Deleuze and Guattari (2014), and Braudel (1982) – and as is specifically evident in Newcastle's own complex post-industrial urbanity and DIY urbanism – the dynamic relations between mainstream and artisanal production are difficult to articulate or reconcile using categorizations that restrict any theorization to a simply binary, linear, or singular ideological model. If we conceptualize the contributory economy and its attendant urbanization according to the polarizing, fluid, and coexistent processes which compose it, we can then acknowledge its strange intertwinement with the same mainstream consumerism and urbanism it productively challenges through its primary focus on social and cultural remediation. It is here, in this site of grassroots-mainstream coexistence characteristic of RN's own DIY urbanism that we see the subversion of conventional depictions of capitalism as a monolithic singularity exclusive of communality and social value in urban settings; a subversion which may emerge from within its own architectures, in a very literal sense.

References

Andrews, B. R. (2009) *Coal, Railways and Mines: The Colliery Railways of the Newcastle District and the Early Coal Shipping Facilities*, vol 1. Redfern NSW: Iron Horse Press.

Augé, M. (1995) *Non-Places: An Introduction to Supermodernity*, trans. J. Howe (2nd ed.), London: Verso.

Barrett, A. (2012) "Five Global Hispster Meccas Even Cooler than Seattle", *The Seattle Globalist* [online], Travel section, April 13, www.seattleglobalist.com/2012/04/13/five-global-hipster-cities-cooler-than-seattle/2296, accessed 13 January 2016.

Bianchi, M. and Maffei, S. (2012) "Could Design Leadership Be Personal? Forecasting New Forms of 'Indie Capitalism'", *Design Management Journal*, vol 7, no 1, pp. 6–17.

Braudel, F. (1982 [1979]). *Civilization and Capitalism 15–18th Century: Volume II: The Wheels of Commerce*, trans S. Reynolds. London: Collins.

Creative Industries Innovation Centre (CIIC) (2010) "Renewing Newcastle: The Post-industrial City of Newcastle Is Reinventing Itself as a Centre for Culture, Art, Music and Crafts Thanks to Initiatives Like Renew Newcastle", *Creative Innovation* [online], Creative Industries Innovation Centre (CIIC). Sydney: University of Technology, http://renewnewcastle.org/media/renewing-newcastle-from-the-creative-industries-innovation-centre/, accessed 13 January 2016.

De Landa, M. (2000) *A Thousand Years of Non-Linear History*, New York: Swerve Editions.

Deleuze, G. and Guattari, F. (2008 [1972]). *Anti-Oedipus: Capitalism and Schizophrenia*, trans. R. Hurley, M. Seem and H. R. Lane. Minneapolis: University of Minnesota Press.

Deleuze, G. and Guattari, F. (2014 [1980]). *A Thousand Plateaus: Capitalism and Schizophrenia*, trans. B. Massumi. London: Bloomsbury.

Deslandes, A. (2013) "Exemplary Amateurism: Thoughts on DIY Urbanism", *Cultural Studies Review*, vol 19, no 3, pp. 216–227.

Finn, D. (2014) "DIY Urbanism: Implications for Cities", *Journal of Urbanism: International Research on Placemaking and Urban Sustainability*, vol 7, no 4, pp. 381–398.

Finney, T. (2012) "Urban Change", *Architecture Australia*, vol 101, no 1, pp. 70–72.

Flanagan, M. and Mitchell, W. (December 2016). An Economic Evaluation of the Renew Newcastle Project: Final report prepared for Renew Newcastle Limited. Newcastle: Centre of Full Employment and Equity (COFFEE), University of Newcastle.

Gibson-Graham, J. K. (2006 [1996]) *The End of Capitalism (As We Knew It)*, Minneapolis: University of Minnesota Press.

Gordon, J. (2014) "High-rise Plan for Newcastle Mall: 19 Storeys on David Jones Car Park", *The Newcastle Herald* [online], Local News section, March, www.theherald.com.au/story/2129631/high-rise-plan-for-newcastle-mall-19-storeys-on-david-jones-car-park/, accessed 30 June 2014.

Harris, M. (2010) "Budding Recovery Starts", *The Herald*, Tuesday 6 April, p. 18.

Harvey, D. (2010) *The Enigma of Capital*, Oxford: Oxford University Press, Inc.

Integrated Design Commission SA (2012) *Economic Benefits of City Activation and Renewal*, Adelaide: Government of South Australia/IDSA.

Iveson, K. (2013) "Cities Within the City: Do-It-Yourself Urbanism and the Right to the City", *International Journal of Urban and Regional Research*, vol 37, no 3, pp. 941–956.

Keily, L. and Shaw, G. (eds) (2013) *People and Place/Coal and Community*, Newcastle: The University of Newcastle Library.

Kendzior, S. (2014) "The Peril of Hipster Economics", *Aljazeera* [online], Opinion (28 May), www.aljazeera.com/indepth/opinion/2014/05/peril-hipster-economics-2014527105521158885.html, accessed 21 July 2015.

Lefebvre, H. (2014) *Toward an Architecture of Enjoyment*, ed. L. Stanek, trans. R. Bononno, Minneapolis: University of Minnesota Press.

Lemmens, P. (2011) "This System Does Not Produce Pleasure Anymore: An Interview with Bernard Stiegler", *Krisis*, no 1, pp. 33–41.

Lindtner, S. (2014) "Hackerspaces and the Internet of Things in China: How Makers Are Reinventing Industrial Production, Innovation, and the Self", *China Information*, vol 28, no 20, pp. 145–167.

Lundy, C. (2012) *History and Becoming: Deleuze's Philosophy of Creativity*, Edinburgh: Edinburgh University Press.

Marshall, K. (2014) "Good Job", *Good Weekend: The Sydney Morning Herald*, 5 July, p. 22.

McCarthy, J. (2010) "Picking Up the Pieces", *Newcastle Herald*, 28 August, p. 4.

Munzner, K. and Shaw, K. (2014) "Renew Who? Benefits and Beneficiaries of Renew Newcastle", *Urban Policy and Research*, vol 33, pp. 1–20.

NSW Government (2014) "Overview", *Urban Growth NSW*, www.urbangrowth.nsw.gov.au/work/projects/newcastle-east-end.aspx, accessed 13 January 2016.

O'Callaghan, J. (2014) "Taking DIY Urbanism to the Classroom", *Archi-Ninja.com*, 'Opinion' section, www.archi-ninja.com/taking-diy-urbanism-to-the-classroom/, accessed 13 April 2014.

Parvin, A. (2013) "Alastair Parvin: Architecture for the People By the People", *TED Talks* [online], 8.35 minutes mp4, February 2013, www.ted.com/talks/alastair_parvin_architecture_for_the_people_by_the_people.html, accessed 5 January 2016.

Ratti, C. and Claudel, M. (2015) *Open Source Architecture*, London: Thames and Hudson.

Reisley, R. and Timpane, J. (2001) *Usonia, New York: Building a Community with Frank Lloyd Wright*, New York: Princeton Architectural Press.

Renew Australia (2015) "About", *Renew Australia*, http://renewaustralia.org/about/, accessed 6 January 2016.

Renew Newcastle (2010) "Renew Newcastle and GPT Win Australia's Leading Business Arts Award", *Renew Newcastle*, 25 October, http://renewnewcastle.org/news/page/41/, accessed 13 January 2016.

Renew Newcastle (2011a) "Project Applications: New Project Application General Information", *Renew Newcastle*, http://renewnewcastle.org/get-involved/project-applications/, accessed 13 January 2016.

Renew Newcastle (2011b) "Partners", *Renew Newcastle*, http://renewnewcastle.org/current-partners/, accessed 13 January 2016.

Renew Newcastle (2016) "FAQ's", *Renew Newcastle*, http://renewnewcastle.org/about/faq/, accessed 13 January 2016.

Smith, C. (2015) "The Artisan, The State and the Binaries of DIY Urbanism", in P. Hogben and J. O'Callaghan (eds) *Proceedings of the Society of Architectural Historians, Australia and New Zealand 32, Architecture, Institutions, Change*, Sydney: SAHANZ, 7–10 July, pp. 616–626.

Stiegler, B. (2010a [2009]) *For a New Critique of Political Economy*, trans. D. Ross, Cambridge: Polity.

Stiegler, B. (2010b) "Bernard Stiegler: Economic Prosperity Relies on Creativity", *Labkulture*, 8 November, www.labkultur.tv/en/blog/interview-bernard-stiegler, accessed 13 November 2014.

Stiegler, B. (2011) *The Decadence of Industrial Democracies: Disbelief and Discredit, Volume 1*, trans. D. Ross. Cambridge: Polity Press.

Stiegler, B. and Fayner, E. (2013) "Bernard Stiegler; 'We are entering an era of contributory work'", in S. Kinsley (trans), *SamKinsley*, pp. 1–5, www.samkinsley.com/2013/02/06/bernard-stiegler-we-are-entering-an-era-of-contributory-work/, accessed 16 December 2013.

Taylor, J. (2011) "DIY Urbanism – Sydney Reconsidered", in L. Stickells and Z. Begg (eds) *The Right To The City*, exhibition catalogue, Sydney: Tin Sheds Gallery, University of Sydney, pp. 46–51.

Westbury, M. (2010) "How Social Media Saved Renew Newcastle", *Marcus Westbury – My Life. On the Internets*, 23 October 2010, www.marcuswestbury.net/2010/10/23/how-social-media-saved-renew-newcastle-vapac-talk/, accessed 6 January 2016.

Westbury, M. (2011) *DIY Transforming a Dying City*, Vimeo video, 8.53min. Newcastle: Renew Newcastle, https://vimeo.com/15759471, accessed 8 April 2014.

Westbury, M. (2013a) "About Marcus", *Marcus Westbury; My life. On the Internets*, 26 October 2013, www.marcuswestbury.net/about, accessed 16 December 2013.

Westbury, M. (2013b) "Cities: Are They Youtube or Hollywood?", *Marcus Westbury; My life. On the Internets*, 22 May 2013, www.marcuswestbury.net/2013/05/22/cities-are-they-youtube-or-hollywood/, accessed 6 January 2016.

Westbury, M. (2015) *Creating Cities*, Melbourne: Niche Press.

Yang, Y.-H. and Lai, I. (2014) *Maker*, DVD, 65 minutes, http://makerthemovie.com/, accessed 8 August 2017.

Zeiger, M. (2011) "The Interventionists Toolkit, Part 1: Places", *Places*, January, https://placesjournal.org/article/the-interventionists-toolkit/, accessed 3 February 2015.

Zeiger, M. (2012) "The Interventionists Toolkit, Part 4: Project, Map, Occupy", *Places*, March, https://placesjournal.org/article/the-interventionists-toolkit-posters-pamphlets-and-guides/, accessed 3 February 2015.

6 Rapid urbanization and Wang Shu's architecture

The use of *spolia* and vernacular traditions in China

Hing-Wah Chau

Introduction

Consumerism has significantly transformed China in the post-Mao era. After the economic reforms of 1978 initiated by Deng Xiaoping, the centrally planned economy of the Mao years (1949–1976) was replaced by a market-oriented economy, which opened China up to overseas investment. Long suppressed entrepreneurial energy was unleashed, rapidly making China one of the world's fastest growing economies, with an average increase in gross domestic product (GDP) of nearly 10 percent from 1979 to 2014 (Morrison 2015, p. 1). Following the 2008 financial crisis, China has more consciously shifted to a consumer society (Gerth 2010, p. 10), reducing its previous over-reliance on exports for economic growth and promoting more domestic consumption. Although China claims to be a socialist country pursuing "socialism with Chinese characteristics," consumerism is now deeply entrenched in the everyday life of the Chinese people (Lin 2006, p. 5).

The logic of consumerism involves the continual process of discarding the old and creating the new for individual satisfaction and more broadly for enhancing economic growth. In fact, consumerism is "the synergistic interaction of mass production and consumption" (Chase 1991, p. 211). In order to generate more sales and create more profits, commodities are designed to wear out faster with built-in obsolescence (Jackson 2009, p. 97). Besides technological obsolescence, various marketing strategies encourage consumers to endlessly seek novel commodities.

Instead of embracing the logic of consumerism, Wang Shu (b.1963) deliberately incorporates the old into his new architecture, especially the creative use of discarded building materials. In doing so, he reiterates the importance of the continuity of living traditions and the preservation of vernacular fabric and its deep connection to local history and culture. Traditional neighborhoods and built environments have been significantly affected by an extensive and accelerated urbanization in contemporary China.

Rapid urbanization

When the People's Republic of China was founded in 1949, there were only 58 cities containing less than 11 percent of the total national population (Campanella

2008, p. 174). Throughout the whole Mao era, less than 20 percent of the Chinese population lived in cities and the urban population even decreased during the Cultural Revolution, as around thirty million urban residents were sent to the countryside for re-education (Campanella 2008, p. 179). At the start of the period of economic reform beginning in 1978, the Chinese urban population was less than 18 percent of the total population. This figure had increased to over 51 percent by 2011 (Cao et al. 2014, p. 412) and is projected to be 70 percent of the total population by 2030, reaching one billion people in total (Denison 2012, p. 52).

Under a relaxation of the once dominant household registration system (*hukou*), people previously living in the countryside are now able to flock to the cities in search of jobs and a better life. A formerly rural population of agricultural laborers has been rapidly converted into an urbanized workforce, bringing about a huge rural-urban migration in China (Edelmann 2008, p. 10), a process that took a century to unfold in the United States (Campanella 2008, p. 281). According to the National Bureau of Statistics in China, 300 million people are expected to move to the cities between 2010 and 2015 alone, more than double Russia's current population (Philips 2013).

Urbanization is the driving force of economic development in China, with abundant labor, cheap land, and established infrastructure. Consequently, the proportion of GDP generated by Chinese cities is expected to increase to over 90 percent by 2025 (McKinsey Global Institute 2009, p. 13). This rapid economic growth has led to the rise of a significant middle class, estimated at around 100 million, including professionals, managers, and private business owners (Hulme 2014, p. 3). The acquisition and possession of new commodities has shaped the identity of this substantial group, and a strong consumer culture has been nurtured among them. This rapidly growing middle class is concentrated in urban areas and exerts a substantial impact on the economy.

In order to stimulate ongoing economic growth and domestic consumption, the Chinese government has been active in planning for this continual expansion of urbanization. However, this unprecedented urbanization in China has occurred at a cost of severe environmental degradation. Due to the expansion of urban sprawl, the total area of farmlands has been significantly reduced to merely 120 million hectares, which is a bare minimum for ensuring food provision and security (Campanella 2008, p. 17). Some of the areas developed in haste have also remained unoccupied, resulting in ghost cities and wasteful real-estate development (World Bank 2014, p. xxiii). The lack of general awareness of the need to preserve natural resources has hindered efforts to strike a balance between economic prosperity and environmental protection.

China's conventional approach to urbanization reflects the influence of globalization. The Western planning model is embraced by adopting common typologies of skyscrapers and shopping malls. These are built very rapidly in huge quantities, transforming and dominating the skyline, supplementing the monumental public squares and wide streets created in the recent past. A homogeneous, modern global cityscape is emerging while regional differences are ignored.

The destruction of traditional built form

Land-intensive urbanization has caused a massive and ongoing destruction of traditional built form in China. The urban landscape in China is changing at dizzying speed: in the midst of rampant construction activities, demolition sites are ubiquitous, generating large amounts of rubble and debris to be disposed of in landfill sites. The number of traditional villages in China has been rapidly and significantly reduced, from 3.6 million in 2000 to 2.7 million in 2010 (Philips 2013). Forty percent of the old Beijing city was razed between 1990 and 2002 for redevelopment (Campanella 2008, p. 150). It has become common for old districts and villages to be demolished for the subsequent construction of commercial skyscrapers, high-end shopping centers, luxury gated communities and residential towers. Such a scale of destruction, involving the demolition of old neighborhoods and displacement of local residents, has led to widespread criticism and resentment.

Facing such a social phenomenon, urban ruins have become a recurrent theme among Chinese artists, such as in the work of Zhang Dali and Yin Xiuzhen. In his *Dialogue* series (1995–2003), Zhang Dali (b. 1963) sprayed graffiti-like silhouettes on half-demolished residential houses with the tag "AK-47," to express the violent social and cultural disruption such mass demolition represents to local residents (Gao 2011, p. 18). Photographs were taken at particular angles to highlight the comparison between walls to be bulldozed and newly constructed iconic landmarks. Even the corner towers of the Forbidden City were used as a backdrop to highlight the contrast with the present experience of ruination. In the work of this artist, the preservation of heritage and the destruction of existing urban fabric are deliberately contrasted with each other. Instead of drawing graffiti, Yin Xiuzhen (b. 1963) collected discarded roof tiles from demolition sites. In the installation, *Airing Tiles* (1998), Yin displayed these roof tiles and attached photos of the demolished houses from where the tiles came (Wu 2012, pp. 208–213). The reused architectural elements of this installation suggest the vanishing traces of everyday life experienced by local residents are now displaced by this relentless destruction.

Wang Shu and the vernacular

In contrast to their artist counterparts, most Chinese architects have been busy immersing themselves in the commercial developments and government projects of the construction boom, without taking any critical stance towards this rapid urban development and its social and material impacts. An exception is Wang Shu, who has attracted widespread international attention, especially after receiving the prestigious Pritzker Prize in 2012. Wang's work was recognized by the Pritzker jury panel as "timeless," with a "strong sense of cultural continuity and re-invigorated tradition" (Pritzker Architecture Prize 2012). Against this background of continuous urbanization and consumerist globalization in China,

Wang emphasizes the lasting importance of connection with locality and regional traditions.

Wang has shown his intense interest in the vernacular fabric of China since he was an architectural student. During his undergraduate studies, he published an article in the Chinese architectural journal, *Architect*, on laneways and narrow streets in old towns (Wang 1984). He expressed his fascination with the delicate scale and proportion of old neighborhoods and urged architects to pay attention to the details of these living environments. He published another, second article during his postgraduate studies, focused on the vernacular architecture of Wannan and its surrounding contexts in Anhui Province (Wang 1987). By thus appreciating the vitality of the vernacular fabric, Wang was criticizing the adverse impacts of the extensive destruction of traditional dwellings and also the erection of new standardized village housing in its place. These two early articles clearly define Wang's design focus on vernacular traditions, which do not merely involve the physical fabric of existing villages and towns, but also the folk practices of its use of materials, its carpentering skills, and surviving craftsmanship.

After graduation, Wang chose not to follow a typical career path and deliberately kept at a distance from official design institutions and private developers. He preferred to work as a freelance designer and even as a construction worker, so that he could equip himself with building skills through interacting with artisans on site. The title of his first monograph, *The Beginning of Design*, reiterates his advocacy of learning through experience and everyday life, and through collaboration with skilled artisans outside the ambit of a globalized professional architectural knowledge (Wang 2002). By naming his atelier "Amateur Architecture Studio," Wang has positioned himself self-consciously as an 'amateur' ready to learn, rather than a professional, and thus global, architect. Against the ethos of his profession, of merely focusing on its work and overlooking the dramatic social contexts of China's transformation, he insists on a self-conscious design practice based on a sensitive observation of the praxis of life, so that an appropriate design can be generated to suit actual needs and local lived contexts (Wang 1994, p. 85).

Wang has also immersed himself in the historical study of traditional, vernacular building practices. In the early 2000s, Wang and his team commenced a series of analyses on folk building construction in Cicheng, in Zhejiang Province. People living in Cicheng are easily affected by typhoons, which can cause their houses to collapse. After natural disasters, local residents often recycle available broken roof tiles and bricks to rebuild their houses within a short period of time. Following his research on Cicheng, discarded tiles and bricks salvaged from demolition sites have been incorporated into his new architecture, and folk building techniques have been applied from the small initial experiment of his Five Scattered Houses in Ningbo (2003–2006), to the larger complex of the Xiangshan Campus of the China Academy of Arts in Hangzhou (2004, 2007), from the Ningbo History Museum (2008), to the urban scale of the Zhongshan Road Revitalization in Hangzhou (2009).

Wang's use of *spolia*

Similar to the recurrent theme of urban ruination and erasure to be found in the work of the Chinese artists described above, Wang clearly claims that ruins are the basis of his work (Wang 2013, p. 46). The reuse of old bricks and tiles can be found in most of his work. As he explained in an interview,

> I like to build with old recycled bricks and tiles in the tradition of the region in which I live. In this way, the materials are saved, various possibilities of material application are expressed, ample and exquisite crafts are developed, and meanings of memory and time are kept.
>
> (Wang in Fuchs 2012, p. 457)

Reused architectural elements and components are commonly known as *spolia*, a word derived from the Latin, *spolium*, which meant originally "the skin or hide of an animal stripped off" or "the arms or armour stripped from a defeated enemy" (Lewis and Short 2005). In its original context, *spolia* referred to trophies seized from enemies and exhibited in triumph. This meaning was then extended to include the reuse of fragments of earlier monuments such as occurred in the Arch of Constantine in Rome (Liverani 2011, p. 45). This reuse of older architectural elements in newer structures has been in evidence in Europe since this early period. *Spolia* is now used extensively to refer to the reuse of various architectural elements in contemporary architecture (Brenk 1987, p. 103).

A prominent example of Wang's use of *spolia* on an international stage was his *Tiles Garden* as the Chinese Pavilion for the Tenth Venice Architecture Biennale (2006). Similar to but larger in scale than Yin Xiuzhen's earlier *Airing Tiles* installation, more than 60,000 old Chinese roof tiles were reused; half of them were laid flat on a bamboo base, while another half were gently sloped on a diagonal line (Wang and Xu 2006, p. 6). *Tiles Garden* was constructed on site in two weeks to demonstrate the actual construction process and working approach of a Chinese architect within time and site constraints. The whole garden was partly bisected by a bamboo ramp that enabled visitors to walk over the roof tiles. Since there is a long tradition of using roof tiles in Chinese vernacular architecture, walking on the bamboo ramp over these tiles suggests symbolically walking over the city, encouraging the visitors to his installation to reflect on the considerable social and cultural impacts of China's rapid urbanization.

The use of *spolia* in Wang's work is not just a matter of adding bricks and tiles, but a deliberate stance against unsustainable consumption embodied in China's dramatic transformation. In the Xiangshan Campus, Wang used more than seven million pieces of old roof tiles and bricks, salvaged from demolition sites. The resulting assemblage does not merely possess decorative value, but itself becomes a visual archive of ruination. Deliberately avoiding any cleaning treatment, vestiges of the past can be seen on the surfaces of these architectural remnants, enabling them to display their original qualities and traces of weathering. In response

Figure 6.1 The use of spolia in Xiangshan Campus Phase II, Hangzhou (2007)

Photo: Hing-wah Chau.

to extensive urbanization and its erasure of local history, the old bricks and tiles of Wang's new architecture becomes "a way of preserving time" (Denison and Ren 2012, pp. 125–126). In the Xiangshan Campus, for example, the original adhesive tapes used on bricks and tiles are deliberatively left untouched, showing their history. This is a material history of an unrecorded vernacular past, of everyday existence with no specific dates or names to celebrate.

Erasure and nostalgia

In addition to displaying traces of the material past, *spolia* can evoke nostalgia, that is "a sentimental longing" for the past (Sedikides et al. 2008, p. 305). As Ritivoi (2002) explains, nostalgia occurs when "the present seems deficient in contrast to the past" and this can arouse an awareness of an individual's personal identity and history, beyond the conditions of the present (pp. 3, 32). Wang does not aim for the "restorative nostalgia" Boym (2001 pp. 41–48) describes, where nostalgia seeks to reconstruct the past itself in memory. Instead, his work evokes a "reflective nostalgia" that raises questions about the role of memory and history to the present (Boym 2001, pp. 49–55). This reflective nostalgia helps counter the erasure of memory and disjuncture with the past that China's urbanization and development has resulted in.

The relationship between memory and place is well elaborated by Paul Connerton in *How Modernity Forgets* (2009). For Connerton (2009, p. 10), memory lies in place-names, the house and the street. Considering consumerism as a way of life based on forgetfulness and obsolescence, he criticizes the scale, production speed, and the "repeated intentional destruction of the built environment," which accelerates the pace of cultural memory loss (Connerton 2009, p. 99). It is important to continue long-standing traditions and maintain the bond between past and present, as the presence of the past can provide a nurturing resource for defining a sense of collective identity (Hutton 2011, p. 98).

In the case of the Ningbo History Museum, the site is described by Wang himself as a "no memory area," because the old villages it replaces have been totally razed for this new urban development (McGetrick 2009, p. 74). However, through the transformation of the rectangular base of the museum into five separate architectural units, Wang deliberately recalls the villages' traditional streetscapes, even maintaining the proportion of the spaces that typically existed between the buildings. Indeed, Wang met a woman who had visited the museum four times within half a year because it awakened her own memory and emotional attachment to what the museum and surrounding development had replaced (Moore 2012, p. 32).

The reflective nostalgia evoked by the Ningbo History Museum is reminiscent of that often associated with David Chipperfield's Neues Museum in Berlin

Figure 6.2 The use of spolia in Ningbo History Museum (2008)
Photo: Hing-wah Chau.

(McGetrick 2009). Traces of decay in the *spolia* used there are also intentionally exposed to encourage reflection on a lost past, an erasure of what had once been. As its architect, Chipperfield wanted to respect both history and place and not "to make a cut with the past" (Emilio 2004, p. 11). Comparable with the Neues Museum, Wang's Ningbo History Museum seems to be a ruin composed of archaeological vestiges. There is a viewing platform on top of the museum providing a panoramic view of the surrounding newly developed area. Viewing the surrounding cityscape in such close proximity to the historical fragments retained on the facades of the museum, can stimulate reflection on the destruction of vernacular fabric the development has entailed, and raise questions about the future of the city.

Indeed, Wang's use of rough weathered materials, in striking contrast to the polished surfaces and pristine cleanliness of the adjacent modern high-rise buildings, provoked criticism from government officials during the construction process. These officials disliked Wang's reuse of discarded artifacts, which seemed to embody a "backwardness" incompatible with the image of the modern metropolis they nurtured (Wang 2009, p. 76). However, Wang's museum was well received after its opening, achieving a satisfaction rate of 97.5 percent from the three million visitors who came to the museum in the first three years after its completion (Hu 2012, p. 206). This suggests that Wang's deliberate evocation of the memory of the past has been widely appreciated.

Figure 6.3 Ningbo History Museum and the surrounding urban context

Photo: Hing-wah Chau.

The use of vernacular traditions

Wang's work is deeply rooted in a strong sense of the necessary continuity of vernacular traditions and their value today. From the first article he published in an architectural journal in the 1980s, to his series of studies in Cicheng town in the 2000s, he shows a persistent concern with the survival of the vernacular fabric and its embodied cultural traditions, and their close relationship to locality, place and nature.

For example, in Wang's Xiangshan Campus, the buildings devoted to teaching are clustered around a central small hill so that the natural environment of half of the site can be preserved and experienced. Between these buildings and the hill, some areas have been maintained as farmlands, and the original brook around the hill has also been left in place on the site. This decision to maintain the natural features of the landscape at the site has had a positive ecological impact. After the completion of the campus, the number of migrating birds that make their seasonal homes there is estimated to have increased from 300 to 3,000 (Tang 2012, p. 56). Deliberately referencing Chinese landscape paintings in his design, Wang emphasizes the central value of a more harmonious relationship between dwelling and nature, implicitly opposing the heavy presence of the manmade in the concrete landscapes of the modern, global city (Muynck 2009, p. 75). In this campus, and in his other work, a more balanced development model with less environmental impact is self-consciously promoted.

In his work, Wang also advocates the application and continuity of folk craftsmanship traditions through his long-term collaboration with local artisans. Building components in his work, such as window hooks, catches, and latches are all made by local ironsmiths. In his exhibition, "Thinking by Hands," held in the Hong Kong and Shenzhen Bi-city Biennale of Urbanism/Architecture in 2007, Wang (2007, p. 45) reiterated the importance of craftsmanship. He drew attention to the reciprocal relations between people and handcrafted artifacts, a relationship that has been undermined by the mass production of commodities. For example, for the timber roof of the Imperial Street Museum in the Zhongshan Road Revitalization project, Hangzhou, a traditional technique for ancient bridge construction was used. Wang and his team spent time to look for appropriate artisans and did repeated experimentation before actual construction began (Wang 2013, p. 109). This persistence shows how an exquisite building skill can be rescued from oblivion and how the continuity of vernacular traditions can be maintained. As Wang claimed, his intention was to "recover the cultural roots that the place had lost" (Denison 2012, p. 124).

Regarding the use of vernacular traditions, Wang is not merely interested in producing a handful of architectural works, but also in offering an alternative design-based approach to urban development in China that addresses regional conditions and the local cultural traditions of the past. Amid rapid urbanization, it is crucial for architects to shift their focus to the fabric of the city itself and how this is conceived (Wang 2007, p. 45). In the Xiangshan Campus, different

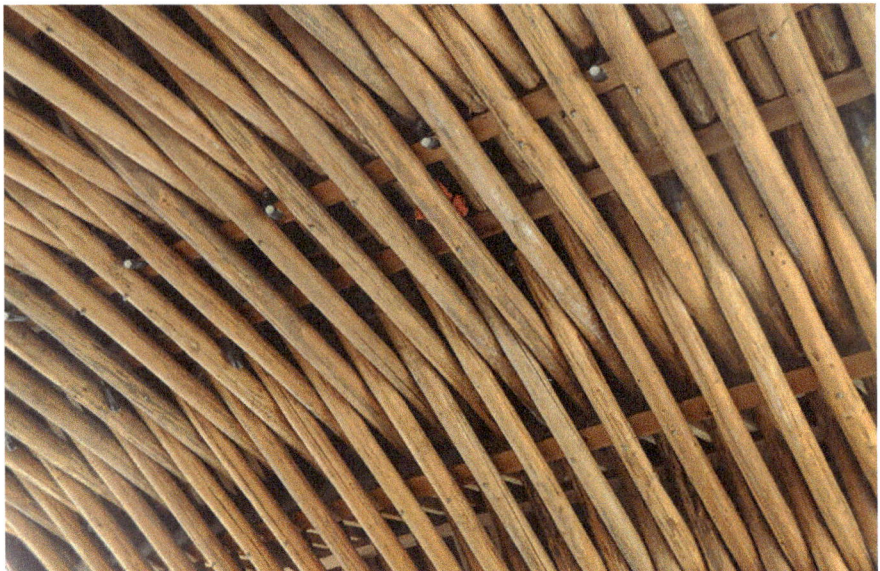

Figure 6.4 Craftsmanship of the timber roof of the Imperial Street Museum, Zhongshan
        Road Revitalisation, Hangzhou

Photo: Hing-wah Chau.

architectural typologies using local materials are arranged in a close-set manner
to establish a highly dense built environment, and one that remains in dialogue
with its natural surroundings. From this perspective, the campus can be seen as an
exemplar of urban design.

Although the Xiangshan Campus is away from the Hangzhou city center,
Wang's design intention is urban focused. He clearly identifies the importance of
continuity in the design of the city's fabric for maintaining the life of community
and city:

> . . . what makes a city interesting is the community it supports, public life,
> and spontaneity. This is why I am opposed to deliberately dismantling the old
> city: it is the core of a city. It is not only about preserving history, it is about
> urban life itself.
>
> (Wang, in Fuchs 2012, p. 459)

The Zhongshan Road Revitalization in Hangzhou demonstrates Wang's opposition to the *tabula rasa* approach common today, which basically demolishes all
existing structures and neglects the existing culture and history of the place in
order to offer a blank canvas for subsequent construction and development. The
old Hangzhou city center along Zhongshan Road has been revitalized without the

Figure 6.5 Xiangshan Campus, Hangzhou
Photo: Hing-wah Chau.

massive destruction of the older buildings typical in most urban developments across China. Although some of the existing buildings were old and dilapidated, most have been preserved and renovated rather than destroyed or left as facades only. Instead of providing wide streets under conventional planning schemes, Wang insisted on reducing the width of Zhongshan Road from the 24 meters recommended by urban planners to 12 meters for maintaining the more traditional proportions of the old town (Magrou 2010, p. 59). The fine-grained urban texture of the site has been respected. As Wang sees it, prevailing planning rules that enforce the creation of wide streets in China are "impossible for the formation of city life" (Wang 2010, p. 109).

Unlike most contemporary profit-oriented commercial redevelopments that also result in the displacement of the area's original inhabitants, the Zhongshan Road Revitalization did not involve any resettlement (Denison 2012, p. 56). A six-month survey was carried out by Wang and his team at the outset to help understand the community in detail (Wang 2013, p. 97). The existing neighborhood shared by the local residents has also been preserved. This revitalization project was not merely about the preservation of a historical precinct, but also focused on the continuity of vernacular traditions and of urban life within it. As Wang has repeatedly emphasized, "regional differences are rooted in living traditions" (Fuchs 2012, p. 454).

Conclusion

Wang's work has become critical in the contemporary Chinese context of radical social transformation. Against global homogenization and overwhelming consumerism, Wang emphasizes the continuity of community and its heritage, of folk building practices, and traditional craftsmanship. By the creative use of *spolia*, his work evokes a "reflective nostalgia" (Boym 2001) for arousing an awareness of the past and its continuing significance. He reiterates the importance of maintaining the bond between past and present and considers the past as a valuable nurturing resource for a distinctive collective identity. Wang's architecture is deeply embedded in the existing contexts that can contribute to cultural, social, and environmental sustainability. Through his work, Wang invites his fellow architects and even the general public to resist the erasure of cultural memory and history generated by China's extensive and destructive urbanization and to reflect on the vernacular traditions and regional specificity in China's rich material heritage.

Acknowledgements

This is derived from my doctoral research (2010–2014) at the University of Melbourne. I am grateful to have received valuable advice from my supervisor, Associate Professor Jianfei Zhu, as well as my panel members, Associate Professor Greg Missingham and Dr. Peter Raisbeck.

References

Boym, S. (2001) *The Future of Nostalgia*, New York: Basic Books.
Brenk, B. (1987) "Spolia from Constantine to Charlemagne: Aesthetics Versus Ideology", *Dumbarton Oaks Papers*, vol 41, pp. 103–109.
Campanella, T. J. (2008) *The Concrete Dragon: China's Urban Revolution and What it Means for the World*, New York: Princeton Architectural Press.
Cao, S., Lv, Y., Zheng, H. and Wang, X. (2014) "Challenges Facing China's Unbalanced Urbanization Strategy", *Land Use Policy*, vol 39, pp. 412–415.
Chase, J. (1991) "The Role of Consumerism in American Architecture", *Journal of Architectural Education*, vol 44, no 4, pp. 211–224.
Connerton, P. (2009) *How Modernity Forgets*, Cambridge: Cambridge University Press.
Denison, E. (2012) "China's Marco-Planning Policies: Architectural Catalyst or Constraint?", *Architectural Design*, vol 82, no 5, pp. 50–57.
Denison, E. and Ren, G. (2012) "The Reluctant Architect: An Interview with Wang Shu of Amateur Architects Studio", *Architectural Design*, vol 82, no 6, pp. 122–129.
Edelmann, F. (ed) (2008) *In the Chinese City: Perspectives on the Transmutations of an Empire*, New York: Actar.
Emilio, T. (2004) "A Conversation with David Chipperfield on a Different Way of Understanding Context", *El Croquis*, no 174/175, pp. 5–25.
Fuchs, C. (2012) "Interview with Wang Shu", *Detail (English Edition)*, vol 5, pp. 454–460.
Gao, M. (2011) *Total Modernity and the Avant-Garde in Twentieth-Century Chinese Art*, Cambridge, MA: MIT Press.

Gerth, K. (2010) *As China Goes, So Goes the World: How Chinese Consumers Are Transforming Wverything*, New York: Hill & Wang.

Hu, J. (2012) "Gongtong Chuangzao Ningbo Bowuguan Jianshe Ceji" ["Together to Create Ningbo Museum Construction"], *Chengshi Huanjing Sheji [Urban Environment Design]*, vol 6, pp. 206–209.

Hulme, A. (2014) *The Changing Landscape of China's Consumerism*, Oxford: Chandos.

Hutton, P. H. (2011) "How the Old Left Has Found a New Place in the Memory Game", *History and Theory*, vol 50, pp. 98–111.

Jackson, T. (2009) *Prosperity Without Growth: Economics for a Finite Planet*, London: Earthscan.

Lewis, C. T. and Short, C. (2005) "Spolium", *A Latin Dictionary*, www.perseus.tufts.edu/hopper/text?doc=Perseus:text:1999.04.0059:entry=spolium, accessed 17 May 2015.

Lin, C. (2006) *The Transformation of Chinese Socialism*, Durham; London: Duke University Press.

Liverani, P. (2011) "Reading Spolia in Late Antiquity and Contemporary Perception", in R. Brilliant and D. Kinney (eds) *Reuse Value: Spolia and Appropriation in Art and Architecture from Constantine to Sherrie Levine*, Burlington, NJ: Ashgate.

Magrou, R. (2010) "Focus on: Wang Shu", *Architecture d'aujourd'hui*, vol 375, pp. 53–91.

McGetrick, B. (2009) "Ningbo Historic Museum", *Domus*, vol 922, pp. 67–75.

McKinsey Global Institute (2009) *Preparing for China Urban Billion*, Shanghai: McKinsey Global Institute, www.mckinsey.com/insights/urbanization/preparing_for_urban_billion_in_china, accessed 18 January 2016.

Moore, R. (2012) "The Quest for Real Life in High-Rise China: Wang Shu's Vision Is a Challenge to His Country's Obsession with Scale", *The Observer*, 16 December 2012, p. 32.

Morrison, W. M. (2015) *China's Economic Rise: History, Trends, Challenges, and Implications for the United States*, Washington, DC: Congressional Research Service.

Muynck, B. (2009) "Local Hero", *Mark*, vol 19, pp. 73–85.

Philips, T. (2013) "China's Villages Vanish Amid Rush for the Cities", www.telegraph.co.uk/news/worldnews/asia/china/10470077/Chinas-villages-vanish-amid-rush-for-the-cities.html, accessed 24 October 2016.

Pritzker Architecture Prize (2012) "Jury Citation", www.pritzkerprize.com/2012/jury-citation, accessed 26 January 2015.

Ritivoi, A. D. (2002) *Yesterday's Self: Nostalgia and the Immigrant Identity*. Lanham, NJ: Rowman & Littlefield.

Sedikides, C., Wildschut, T., Arndt, J. and Routledge, C. (2008) "Nostalgia: Past, Present, and Future", *Current Directions in Psychological Science*, vol 17, no 5, pp. 304–307.

Tang, Y. (2012) "An Architect Respectful of Traditions and Nature", *China Today*, vol 61, no 6, pp. 54–56.

Wang, S. (1984) "Jiuchengzhen Shangye Jiefang yu Juzhu Linong de Shenghuo Huanjing" ["Commercial Districts of Old Towns and the Living Environment of Laneways"], *Jianzhushi [Architect]*, vol 18, pp. 104–112.

Wang, S. (1987) "Wannan Cunzhen Xiangdao de Neijiegou Jiexi" ["Analysis of the Inner Structure of the Laneways in the Wannan Vernacular Architecture"], *Jianzhushi [Architect]*, vol 28, pp. 62–66.

Wang, S. (1994) "Kongjian Shihua: Liangze Jianzhu Sheji Xizuo de Chuangzuo Shouji" ["Space Poetry: Creation Manuscripts of Two Architectural Designs"], *Jianzhushi [Architect]*, vol 61, pp. 85–92.

Wang, S. (2002) *Sheji de Kaishi [The Beginning of Design]*, Beijing: Zhongguo Jianzhu Gongye Chubanshe.

Wang, S. (2007) "'Thinking by Hands' An Experiment on the Reconstruction of Living Places in Collapsing Traditional Cities", in The Hong Kong Institute of Architects (ed) *Refabricating City: Hong Kong & Shenzhen Bi-city Biennale of Urbanism/Architecture*, Hong Kong: The Hong Kong Institute of Architects, p. 45.

Wang, S. (2009) "Ziran Xingtai de Xushi yu Jihe: Ningbo Bumuguan Chuangzuo Biji" ["The Narration and Geometry of Natural Appearance: Notes on the Design of Ningbo Historical Museum"], *Shidai Jianzhu [Time + Architecture]*, vol 3, pp. 66–79.

Wang, S. (2010) "Zhongshan Road", *Lotus International*, vol 141, pp. 106–111.

Wang, S. (2013) *Building a Different World in Accordance with Principles of Nature*, Paris: Editions des Cendres.

Wang, S and Xu, J. (2006) *Tiles Garden: A Dialogue Beyond City, Between an Architect and an Artist (China Pavilion – 10th International Architecture Exhibition, La Biennale di Venezia, 2006)*, Hangzhou: China Academy of Art.

World Bank (2014) *Urban China: Toward Efficient, Inclusive, and Sustainable Urbanization*, Washington, DC: World Bank Group.

Wu, H. (2012) *A Story of Ruins: Presence and Absence in Chinese Art and Visual Culture*, Princeton: Princeton University Press.

7 Public space for changing times

Reuse strategies in transforming the 'wastelands' of cities

Gini Lee

Introduction: public wasteland transformation as a spatial design practice

Drawing upon a range of definitions for 'waste' in the environment beyond simply that deriving from objectified 'waste' forms and materials, this chapter proposes an expanded spatial concept for 'wasted', overlooked, or disregarded places on the margins, or for sites in various states of decay or misuse. Once regarded as the spatial waste products of development, destruction, and/or decline, many contemporary revitalized public spaces display a range of design approaches that have influenced the renewal of landscapes to become fit for creative, safe, and enduring occupation. Investigations into reuse programs for spatially wasted sites reveal a variety of design strategies adopted for postproducing landscapes, alongside an identifiable design lexicon for waste concepts to inform practical tactics for urban transformation. By way of explanation, selected sites that demonstrate various types of spatial wasteland conditions and strategies are examined below; a post-Olympic alternative London, urban space curation in a Cape Town township, and post-earthquake Christchurch. Each of these places is subject to either uninvited change or continuing decline, and each, through what might broadly be understood as design intent, has developed programs for reactivation through physically subtle yet socially engaging spatial invention. Critical economic, political, and natural events have produced new forms of lived space and experience, with intentional and unintentional consequences. To underpin perspectives on the language of waste as space, a review of various contemporary landscape writings on the nature of waste and the waste landscape is included to inform the aforementioned design lexicon.

A design conservationist program for modes of changing occupation in the public realm works towards renewal of places through acknowledging and building upon a site's existing topographical, material, and cultural qualities. Tactics such as the appropriation and repurposing of existing models of space and event are critical to the future imagining and reuse of public spaces. Bourriaud's (2002, p. 12) postproduction theory proposes a recycling practice (for artists) where "works are created on the basis of pre-existing works . . ." and that these works "contribute to the eradication of the traditional distinction between production and consumption, creation and copy, readymade and original work." Postproduction approaches inform redesign practice for disused and/or damaged urban sites through applying physical reinvention in concert with making use of the social and technological networks available to reactivate them. This combination of spatial re-envisioning and event programming can promote public places responsive to the palimpsest of everyday life inscribed on the site over time.

The adoption of postproduction theory as an aid to imagining and managing present and future landscapes is juxtaposed against the often-confronting realities of the transformation of affected places. The resulting conclusions seek to uncover useful strategies for reuse in the face of difficult or calamitous events in challenging post-urban landscapes. The following example drawn from art practice typifies where spatial transformation over time produces unsettling social, environmental, and economic conditions, underpinning the emergence of spatial waste(lands).

On the nature of reuse, erasure, and wastelands: Kentridge's Johannesburg

The ambiguous and transformative possibilities that arise from operating with marginal, overused, or undeveloped landscapes and places described as waste(d) space resonates in the creative work of the South African artist William Kentridge. He recounts the case of the destruction of the gold mining sites that created an entirely new topography in the plains or high veld landscape of the Witwatersrand, where his native Johannesburg was built around the South African gold rush of the late 1800s (Kentridge 2014). Kentridge's films and drawings recount the present-day situation where the artificial hills that are the resulting spoil from the mines have been appropriated through grassing and planting over the past 100 or so years, becoming in turn an essential character of the area's landscape and cityscape. They provide a setting where trees and an artificial nature enable some protection from what was regarded as the unrelenting flat and dry grassy plains of the veld. Yet these actions have provided the city with its definable industrial, albeit artificial, identity. It is in these original post-mining areas where many informal and poor settlements have been established adjacent the landscaped spoils of mining waste. In a twenty-first century spirit of opportunity wrought by technological advances, these iconic hills are now being turned over – reused – to extract every remaining gram of gold in a new type of reworking gold rush.

The artist's primary tools are paper, charcoal, an eraser, and animation, and his works chart the malleability of the landscape in a process of drawing and erasing over and over. Mark Gevisser (2014) writes on his admiration for William Kentridge as he discusses the idea of the mutable landscape of "perpetual sketching and erasing, building and modification." He recounts moments from the film *Other Faces* (2011), where mine dumps were rubbed over in a process that dissolved the certainty of the landscape for the city's inhabitants. Where Kentridge had always thought of the hills as permanent and completely of their place, to see this shift towards nothing produces a feeling of profound dislocation (Gevisser

2014). One can only imagine how these processes affect the private and public lives of those existing often in abject conditions around the mine hills that are in reality dumps to be re-mined, resulting in a flattening of the landscape returning to a facsimile of its earlier self. Undoubtedly the tracks, streets, and public spaces that emerge between the informal dwellings, and the areas where commerce and public life is played out, are equally regarded as marginal to the agendas of the re-miners.

The reuse of landscape type and space becomes a palimpsest, where the fabric of society, history, economics, and transient topography is revealed through seeking to understand what lies beneath the surface of things, lives, and forms. These conditions are shaped not simply through the exploitation of the native landscape, but are also open to re-looking, re-working, and repurposing through economic, social, political, and environmental local and global drivers. Landscape is never in stasis. Usefully responsive techniques for change, driven by transformative circumstances, often result in contradictory outcomes for environment and people – for public benefit and yet also to the detriment of both.

Public space, postproduction, and zones of activity

Various forms of public and private space are produced where people occupy dynamic 'waste' landscapes, and their shape, space, and texture is usually culturally framed and frequently acted upon by challenging environmental forces. The diverse scale and form of such poorly regarded public space is situated across territories, from the urban center to the margins and beyond. No landscape can be left untouched by circumstance. To effect dynamic and useful change to these marginalized places, appropriation and repurposing of existing models of space and event is critical. The future reuse of public landscapes is open to negotiation and realization through novel design approaches to reactivate spaces for the public good. By taking on board the concept of postproduction as a recycling practice applied to urban places and networks of reinvention, it is possible to demonstrate how re-envisioning existing spaces, through reuse of their spatial and curatorial programs, can make resilient, mobile, and economically valued everyday places anew.

Such discursive situations provide opportunities for novel and serendipitous curatorial programs where multidimensional relationships are free to operate between the animate and the inanimate in public space. Recently, design practices such as muf architects' *Hackney Wick Arts Strategy* in London (muf 2014), and event and occupation programs such as *Parklets* in San Francisco (Pavement to Parks 2015), have repurposed existing public space through the inventive curation of temporary events with minimal structural intervention to the site. These approaches acknowledge that urban strategies employ network systems to curate space and events across the fabric of the city beyond the specificity of the site. Further, the concept of networked space expands beyond city limits where the desire for public realm opportunities interrupts the industrial search for fallow land upon which to deposit the refuse of other activities, whether they be mining, dumping, or retrofitting. Such networks spread beyond the public realm through increasing conglomeration of events and/or installations, some sponsored by big business in an effort to improve their public persona while simultaneously reducing their vulnerability to costly remediation and making-good processes.

The common denominator in reused space, whether urban or on the fringes, is that land is regarded as inactive, substandard wasteland, whether it be inner city

Figure 7.1 Istanbul Markets Closed (2002)

Photo: Gini Lee.

streets and vacant lots or the desert(ed) spaces of the periphery. Postproduction theory nominates conceptualizing areas of activity as open-ended zones that allow the insertion of new uses into existing places and situations, founded upon the stuff that is locally and readily available for reuse (Bourriaud 2002, pp. 10–18). Zones of activity enable the programming of existing situations and forms by inserting new activities and systems which may be temporary or transient, but also may endure over time, as event forms rather than physical compositions or structures. An exemplar of this form is the flea and fruit market that arrives without notice in spaces programmed for other uses.

I consider the flea market as a zone of activity and as a place of endless opportunity, where ordered disorder is the event form and the street is designated as a multivalent public space. My visit to a morning street market in Istanbul recalled the dynamic nature of temporary occupation through its daily relocation of simple wooden trestle tables from the street to boundary wall and back again, day after day, to convey the open or closed condition of both street and market. Spatial appropriation in concert with the socio-economic market event program characterizes the tactical aspect of postproduction in practice (Lee 2006, p. 16).

Postproduction, like the flea market, allows for a way of understanding the temporal, and the potential, in the abject spaces that are regarded as a waste of space and time, but also simultaneously as usable spaces for some other activity. Bourriaud (2002, p. 51) suggests a number of tactics towards facilitating the reuse of waste space, which are paraphrased as pertinent to reading existing design works in the public realm and to proposing strategies for implementation:

> *Appropriation:* of the everyday object and those other forms which are readily available for reuse involving selection and modification of that which already exists for insertion into new scenarios;
>
> *Mobility and nomadism:* to bring together the notion of ephemeral community where new nomadic identities can be adopted while traversing and performing in public spaces forming zones of activity for temporary inhabitation;
>
> *Narrative function:* that extends and reinterprets preceding narratives of place to both challenge and invite the passive consumer to engage in collaboration and negotiation while traversing space and time;
>
> *Recycling/sampling/editing:* of sounds, images, forms, and artefacts generating site/locations where already produced forms drawn from cultural history (and nature) are navigated in new ways; and
>
> *Montage:* multi-media space and object are composed by assembling, overlaying, and overlapping different materials, media and/or actors, or curating a series of dissolves, superimpositions or cuts used to condense time or to suggest memories.

In the zone of activity embodied in the temporary market, these tactics are present in a variety of ways: appropriation of the street space and its programming, mobility of borrowed objects – the tables and trestles – alongside the ready availability of produce and people, the narrative function of the tradition of street markets in this ancient city over time, the recycling and editing of the material of the market drawn from elsewhere and repositioned to form the market spatially in the street, and finally, montage scenarios during the time of the market where the objects and materials on display for sale are interpreted by the interactions between seller and buyer. The challenge is to translate these more performative and abstract tactics into a 'designerly' conceptual renewal of marginal waste spaces. Designers typically see these situations as opportunities beyond the spatial appropriation of simple props, requiring substantial infrastructural intervention to make good and/or to make anew.

On the nature of waste and space: reuse of what and where?

A discursive language for the spatially defined waste landscape emerges in fragments from writings by geographical and cultural commentators. A lexicon for waste and place includes phrases such as waste(d)land, waste of space, laid waste, places of last resort, and so on. Writers such as Kevin Lynch (1990) and Grady Clay (1994) promote language that evokes the periphery, the remote, the marginal, and the transgressive, thus releasing waste landscapes from a need to conform to the everyday ideal. Such places may be regarded as zones of otherness or even liberation, enabling the conceptualizing of new zones of activity where opportunities for activity and event arise in the altered and expressive forms of boundary zones, disturbed places, ruined landscapes, and "empty" plains surfaces (Engler 2004, pp. 36–37).

Mira Engler (2004) is largely concerned with waste in the landscape as object located in urban and peri-urban domains. She proposes typologies for specific places of production formed by and servicing the waste stream. For Engler (2004), waste is generally placed in marginal places, which in turn confirms the regard for these landscapes as "inherently inferior sites"; public waste dumps, recycle reuse centers, and sewage facilities. She suggests that these waste sites could be the new sublime, possessing wilderness-like, and thus unoccupiable, qualities of intangible atmospheres and transgressive materiality. Yet she also provides a designer's approach to concepts of spatial reuse for worthy social and environmental programs, albeit problematically often located on contaminated land (Engler 2004, p. xvii).

Citing Robert Smithson's (1967) words, "the conflict between waste and society rests in aesthetics and the imagination," Engler (2004, p. xvii) suggests we have much to learn from environmental art practitioners and writers who have been drawn to marginal landscapes that exhibit the visible attributes of wastelands. Seeking inspiration from the materiality and form of damaged places, Smithson, and others are intrigued by the aesthetics and beauty of sites of waste or sites that appear empty, barren, or uninhabited, such as to be found in the unrelenting wilderness of places not deemed picturesque. Landscape architect Elizabeth Meyer (2008, p. 8) writes on sustaining beauty in the face of the appearance of environmental degradation in relation to expanding perceptions of sustainable design, citing the beauty of the toxic water waste in AMD Park, Pennsylvania, as a "hybrid language of description and aesthetic appreciation." And through examining Smithson's built projects for a disused quarry in the Netherlands, *Broken Circle* (1971) and *Spiral Hills* (1971), Engler (2004, p. 29) suggests that these projects demonstrate the inherent ambiguity and open-endedness of marginal places, where the intent to remake them tunes their ruined/wasted qualities into new and more conspicuous forms. In contrast to typical rehabilitation processes that seek to "reverse or arrest time" into a seamless before-situation, so often resulting in nostalgic pastiche, celebrating waste space aesthetics expresses opportunity for design strategies to work with the tenet that "waste is in the mind as much as reality" (Engler 2004, p. 14).

Engler (2004, pp. 36–40) also posits a proactive language for approaching the typology of landscapes of waste; camouflage as the aesthetics of disguise, utilitarian recycling into places of amenity, restoration to ameliorate disturbed places into a simulated historic condition, and mitigation through technological adaptation. Through acknowledging the complexity and dynamics of sites, she advocates for design-based concepts that operate with the utilitarian, the experimental, the participatory, and the representational. More recently, design approaches have taken a curatorial perspective through educative, experiential public awareness programs and celebrative art and design events intended to expose the community to ongoing transformative processes and effects. Strategies for sustainable outcomes for reworking industrial processing sites adopt mechanisms based upon conservation principles underpinned by economic concerns and self-sufficient programs. Latz and Partners' (1991) well known *Landschaftspark Duisburg-Nord* in Germany's Ruhr region is a past coal and steel industrial site designed to make use of the pre-existing constructed fabric in an otherwise ruined site. Subsequently becoming the model for post-industrial site redevelopment around the world, Duisberg Nord appropriates industrial infrastructure and the narratives of past occupation through redesign and curation of abandoned space and form.

Alan Berger (2006) posits an alternative perspective to this in his reading of the landscapes and patterns of the American suburb in a condition he describes as "drosscape," while revealing the unsightliness and the beauty of these ostensibly unaesthetic places. Berger looks at concepts for waste landscapes in order to suggest typologies of form, to develop a definition for drosscape and ultimately a proposal for drosscape design strategies leading to interventions. He posits various concepts for defining this condition. Firstly, by defining "dross, or waste as landscapes 'scaped', resurfaced, and reprogrammed for adaptive reuse" he nominates dross as a consequence of primary processes resulting from past regimes of urban and industrial production (Berger 2006, p. 12). His investigations encompass the entire urbanized landscape where the drosscape is realized through urban planning patterns promoting sprawl, with aerial, rather than ground-based surveys informing this oversight. Here, the waste product is the overutilization of the landscape in the process of making new suburbs. The vectors of these compositions of wasteful use are the houses, roads, and bridges, industrial and commercial sites, all too often fallen into disrepair and obsolescence. For Berger (2006, p. 12) these drosscapes are sites of potential, his intent being "to identify waste landscapes for future recycling." Mapping typical American urban and suburban situations enable further definitions to emerge:

Drosscape (noun) is the condition of vast, wasted or wasteful land surfaces remade to new sets of values that remove or replace real or perceived wasteful aspects of geographical space.

(Berger 2006, pp. 236–237)

Drosscaping (verb) involves placement upon the landscape of new social programs that transform waste (real or perceived) into more productive urbanised landscapes under programs that require strategic phasing over time.

(Berger 2006, pp. 236–237)

Berger's manifesto (2006, p. 14) extols "drosscape (as) a term for a new design paradigm that emphasizes the productive integration and reuse of waste landscapes in the urban world." He regards dross as a natural urban dynamic accumulating through processes of technological and industrial change, resulting in interstitial, left-over space open to adaptation and reoccupation depending on ground conditions, yet often lacking contextual precedence for aesthetic improvement. Berger calls upon designers to act as expert and advocate for drosscape amelioration and to operate as both spokesperson for and integrator of left-over waste landscapes. As political agents of change in the transformation of unsightly places into more visually pleasing and usefully operable landscapes, designers are encouraged to support programs based upon "choreographing the time/space relationships of reuse projects, alongside adopting a curatorial perspective on the 'productive reintegration' of waste material" (Berger 2006, p. 237). Indeed, Berger (2006, pp. 140–220) is one of the few writers to promote authority for design in what is often a technologically and engineering led recovery, advocating bottom-up approaches as important strategies to counter the top down projects often associated with large scale redevelopment.

Moving beyond descriptive framing typologies of wasted space scenarios, where might the opportunity for the generation of public space for community, recreation, health, and the related public choreography of creative endeavors lie? Reuse strategies need to be cognizant of the tension between the physical and often unproductive realm of the spatial city now overlain by social network technologies and programs that promote urban inconsistency, abstraction, and mobility across cultures and environments.

Alberto Bertagna and Sara Marini's (2011) edited publication *The Landscape of Waste*, curates writings by architects, planners, designers, and cultural theorists who propose theoretical and multi-layered concepts for approaching, understanding, and managing landscapes that exhibit attributes of material and spatial waste conditions. In this collection, Enrico Fontanari (2011, p. 96) focuses on the objects of waste as proponents for zones of activity of resistance/resilience in transformative projects that retain the existence of the underlying site. Hoping to return to a past order, he sees these programs adopting recycling concepts rather than seeking to regenerate former conditions. He makes the distinction that recycling can lead to new arrangements of objects to produce a transformed landscape that nonetheless harbors evidence of previous incarnations and values that may be symbolic, natural, and/or predicated on original use, all exhibiting forms of memory value as a palimpsest. The critical work of the designer resides in the development of criteria to "modify the arrangement of objects" with the attendant necessity of creating a transient boundary to place limits on the transformative program (Fontanari 2011, p. 96). As a project for creating cultural awareness of the importance of the archaeology of landscapes of waste, his curatorial perspective regards such works as object rearrangements that are then represented to the communities from which the 'waste' first originated.

If we seek to look at 'waste' differently, as a transformative agent, then this "forces us to think, to create new objects, new realities and relations, unexpected landscapes," and to develop tactics such as "recycling, juxtaposing and hybridizing" (Fontanari 2011, p. 97). The fragmented structures of landscapes of waste are difficult to strategize as they operate across physical, commercial, environmental, and political processes. As spatially fluid sites and domains, these landscapes require new references, new understandings of porous boundaries, and new spatial practices under regimes that support open-ended and relational enquiry, novel design practice, and a deep knowledge of physical and intangible systems over time. Informality is the new normal in the disaggregated and networked contemporary city.

Using the complexity of the place of Venice as inspiration, guide, and exemplar, its territory and the multiplicity of conditions that make it an enduring yet fragile place, another contributor to this collection, Renato Bocchi (2011, p. 52), writes poetically of the waste 'land-scape' (his inflection) of the islands and the lagoon from an aesthetic of "experience, atmosphere and transfiguration" with water as the connector. He promotes the idea that the lagoon can educate through returning to the materiality of site conditions and dynamics, proposing that "waterscapes may restore the identity of the landscape" (Bocchi 2011, p. 52). His design process draws upon literary and artistic sources to allude to processes of re-composition, of complementarity, and of integrated parts while adopting tactics of montage to utilize spatial and narrative fragments drawn from the landscape as "a palimpsest that is continuously erased and rewritten, . . . where traces remain to build a continuum; traces of culture, the geo-archaeological layers that are our heritage," and erasure is kept to a minimum (Bocchi 2011, pp. 56–57).

Although as a mid-twentieth century artist, Robert Smithson (1967) predates Bourriaud's theoretical investigations, he could well be regarded as an adherent and practitioner of postproduction techniques. He evoked ways of experiencing ruined spaces as monuments and ascribing value to disregarded places through provocative new perspectives on the marginalized, demonstrated by his installations and writings. Smithson's (1967) stories of walks through the "Monuments of

Passaic, New Jersey" remain a definitive account of rereading (waste) landscapes through reinterpretation while touring,

> building a montage of things and pictures capable of narrating in a simultane-
> ous and continuously evolving spatial framework rather than a chronological
> sequence; a palimpsest that is continuously erased and rewritten, but where
> traces remain to build a continuum; traces of culture, the geo-archaeological
> layers that are our heritage.
>
> (Bocchi 2011, p. 57)

This expands upon the understanding that the waste landscape is a palimpsest to be reconstructed and reworked through engagement with traces and forms of the past, to be further transformed into a new order, and then recycled back into altered relationships with past fragments (Bocchi 2011, p. 59). Beyond the narrative and the tour, reconfigured waste spaces utilize and defend the ephemeral alongside the physical and material qualities of places. Italo Calvino's (1974) *Invisible Cities*, arguably one of the most poetic and mysterious works on the phenomenon that is Venice, employs fragments built up over time to retrace and to conjure up new realties and myths, thus adding to the palimpsest that is the essence of this city, a place undergoing physical ruination from both environmental forces and the attendant problems that go hand-in-hand with too much social success. Venice today exists (and ever has) as both wasteland and as a future place, in a state of constant mutability over time as the city and its environs are reconceptualized and reworked by successive regimes to hold back the wasting processes of time.

Forms of waste as space in practice

To return to the language and realities of spaces and places that may be regarded as waste(d) but are now transformed as reuse zones of activity, three examples follow where novel design intent and intervention has enabled transformations that produce new narratives and occupations. Experienced as spaces where reuse has been achieved or is currently in play, the following is a reading of these projects through a postproduction lens to clarify design strategy and practice actions undertaken. While each place exhibits multiple qualities, the review is structured to highlight a particular dynamic at play in the public urban landscape: 'Waste of Space' reused as meeting and performance space for art, 'Wasteland' repurposed for safe travel and 'Laid Waste' as urban reconstruction on a ruined landscape. The following narratives recount personal visits made to the places selected, to seek an understanding of the forces and forms produced through what might be termed postproduction processes.

Waste of space – street interrupted: Hackney Wick post-Olympics 2012

There is a distinct contrast between the old and the new across the eastern part of London, recently re-envisioned for the 2012 Olympic Games, and for the life of the city beyond that event. Wandering through the Queen Victoria Gardens,

Figure 7.2 Hackney Wick, street repurposing where the café meets the street (2012)
Photo: Gini Lee.

designed and detailed for crowds on land that was deemed marginal enough to build a vast Olympic venue, the urban landscape here is an assemblage of parks, gardens, and paved or graveled spaces that are now an open space legacy of the "green" games of London. Further west, Hackney Wick and Fish Island display the conglomeration of structures and spaces that seemingly remain undisturbed by nearby international events. This place is entered by crossing a narrow bridge pathway over a canal of houseboats and barges into a world of curious juxtapositions, and the material signals of creative forces at work among the detritus of post-industrial England. Not far into the streets, laneways, and assorted building types, an outdoor café space appears, occupying the space of a functioning roadway, furnished with a collection of oddly mismatched furniture and grimy urban details. It was more inviting than the emptied-out spaces of the large parks, and it was inhabited, albeit sparsely, on one of those cold London days where nobody ventures out unless absolutely necessary.

This space is described as the Olympic Fringe in planning and design documents, where it is difficult to assess whether the fringe descriptor is regarded as pejorative or as edgy opportunity. Muf architecture/art and J&L Gibbons were approached by the local boroughs and other private and public developers to find novel ways to improve the fabric, economic, and social qualities of the area, located now in the shadow of the Games sites. Their response was to develop design strategies for a network of local and diverse places, where interventions could occur through community action on small, overlooked, and/or spaces reserved for other uses (muf 2010). muf and J&L Gibbons describe aspects of the urban landscapes that are encountered here as 'ruderal' (muf 2014), otherwise defined as growing in trash. The locals proposed an Art Manifesto (muf 2014) "to develop the necessary conditions for creativity" such as the visible and productive presence of artists, an artistic urban character cognizant of historic fabric, even in its 'run-down' state, small scale, of benefit to the public, traffic free, advocating acts of undercover resistance, and finally mapped and recorded for its own legacy project.

Hackney Wick has become a zone of activity through this activation of public space through the adoption of three strategic themes: "Green Infrastructure," "Ways (With) In,' and "Olympic Face/Local Amenity." These themes were effected through adopting a programming tactic for a series of 'home grown' art events played out in public spaces, which aimed to collect diverse and disparate community and social groups to work together on a series of 21 small projects (muf 2014). The designers saw that the physical network of streets, canals, and left-out or left-over space fragments could support a network of art and design projects for the public realm. Their projects are about connectivity and expression and appeal to the inherent mobility of the east London art community, and the material and heritage narratives that interpret the sense of the ragbag fabric of the place. Everyday forms and materials are appropriated to create new temporary spaces that may become permanent in time. Even the master plan exudes a light touch, combining spatial structure to scale and graphic narratives that seek to convey the tangible spirit of each place, through a curated montage of space, material, and event.

Waste land – Khayelitsha Urban renewal 2011

When social engineering meets top down planning on ecologically difficult terrain, the legacy of poor development for disadvantaged communities results in a loss of security, unsafe living environments, and a lack of community interaction in poor quality urban places. A potent example of advocating for careful design strategies to focus on the basic elements of safe passage in the public realm is demonstrated in Khayelitsha, a township in Cape Town. This is a groundbreaking contemporary landscape project by Tarna Klitzner that operates beyond the mainstream, seeking to remake existing yet dangerous use patterns into a journey of safe passage. The scope of the aspiration presented by the reworking of a simple couple of kilometers walk from the square to the railways station is commendable. This walk traverses a sequence of domestic, recreational, educational, and commercial spaces that are the life of the town. The pathway acts as a desire line between places – the shortest path between destinations – and follows the original narrow pathway between house, shops, yards, and untended open space; dark at night, windswept, and hot during the day. The pathway bustles with life and all feels safe, yet shepherds and guides are a necessary aspect of the surveillance system in place to support safe passage along the route between spaces known as safe nodes, where new buildings and public amenities support peoples' confidence to leave their houses and join the community outside.

Historically, apartheid planning saw the township built over the challenging Cape Flat landscape of high water table and propensity for winter flooding. Jared Green (2012) relates that the detention basins formed to manage winter wetland overflows were once the preferred place for dumping bodies, even though these basins are beside the desire path, which everyone must use. Funded by the city, development banks and local not for profits, the Violence Prevention through Urban Upgrading program (VPUU) has contributed to lowering crime rates in this area by a substantial percentage (Umusama 2015).

Once a spatial waste land, the newly designed and paved pathway articulates zones of activity as drivers for safe passage such as the original detention basins, which have now been turned into sports areas and still function for water management when required, and also serve as spaces for education and community meetings. Platforms for seating and gathering alongside the development of water tap areas – for many their only source of water – function as communal places with small infrastructures, where once there was merely a simple standpipe draining into the ground. Although the dry climate and barren soils make it almost impossible to establish trees, each water place sponsors its own tree through proximity to the tap, helped along by the overflows that occur from time to time during water gathering. Such simplicity in companion planting, community space, and

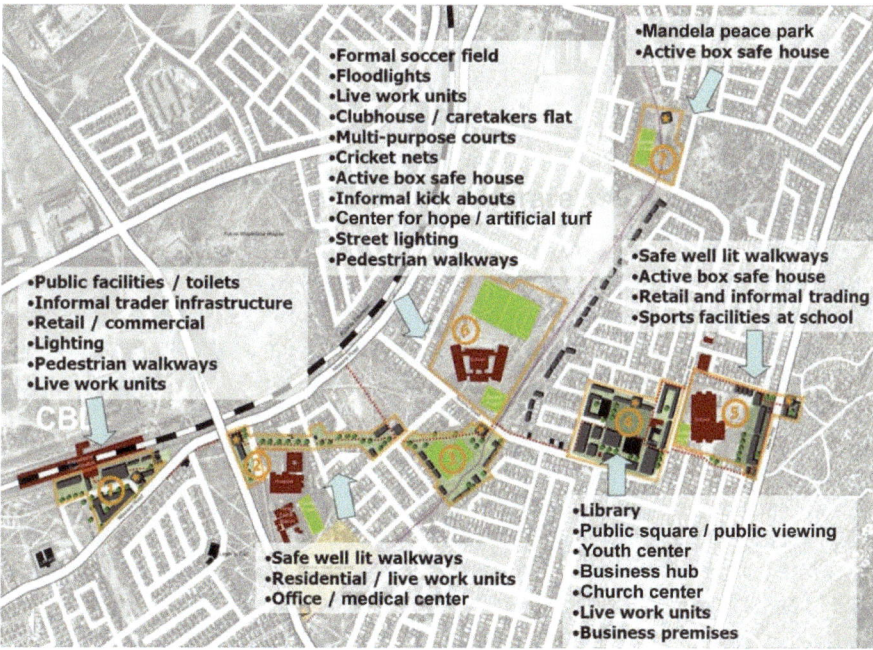

Figure 7.3 The walk from library and square to the railways station, Khayelitsha[1]

Figure 7.4 Walking from the Square to the train station via safe passage. Tarna Klitzner,
          Landscape Architect (2014)

Photo: Gini Lee.

water supply, provides for community programs that encourage children to learn
to tend trees, and to encourage new knowledge of the environment that supports
their homes and economies. Each small zone of activity contributes to the greater
network, with a scale that is both local and human-centered, and in a form based
upon mapping relationships between public and private urban space, and between
social and antisocial behavior, in recognition of the high degree of mobility of
individuals within the community. Newly structured space, with clearly defined
boundaries to antisocial behavior, no longer supported by poorly designed and
maintained urban space, is central to the VPUU framework, which has been built
upon the writings of criminologist C. Ray Jeffery (1977) in *Crime Prevention
through Environmental Design*.

   Transformations in public space behavior, such as that demonstrated in Khayelitsha, helps develop an understanding of the multifaceted conditions that inform the
tactical reuse of what was once regarded as wasteland. The township has resulted
from a poverty-driven social and economic situation, exacerbated by a combination
of landscape features, of geology, topography, climate, and hydrology, to provide
extremely challenging environmental conditions. The design tactics employed for
unmaking/repurposing such left-over space to enable the hoped-for safe passage

have appropriated the existing desire path and remade it in an expanded form by providing a structural layer over the original. The path kinks and adjusts according to the pre-existing plan and pauses as spaces open out or close according to existing narrative forms, aptly illustrated by the various handmade advertising signs attached to house, shop, or shed. The project works with the drift lines and the story lines of the people and their activities – buying and selling, learning and playing, public and private, and adjusting to security from danger towards a sense of community and public realm, activated by the superimposition of local forms and practices.

Laid waste – Christchurch post-earthquakes 2011 to 2015

This southern New Zealand city was devastated – laid waste – by a series of major earthquakes and aftershocks from 2010 to 2011, with small shocks continuing some years later. The city as people had known it disappeared in hours, with the commercial center and areas around the river and floodplain reduced to rubble and watery, sandy mush. Revealed from above by aerial and satellite imagery, the patterns of what remains are compelling. Everything built collapses or sinks into liquefied silt, but the appearance of green suggests that natural places such as the Avon River, the gardens, and the trees apparently endure. The destruction was so extreme that new ways of describing the city arose, with zones of activity renamed according to the degree of wasteness; the 'Red Zone' – the complete gridded center of the city – is a fenced no-go zone, along with the suburban zones along the river flats, and is marked for demolition and greening; the 'Amber Zone' exhibits more abstract patterning between ruination and remnant structures; and the 'Green Zone' is regarded as inhabitable.

Among many issues in the altered city was that the public spaces of the city also disappeared. The newly mobile people of Christchurch were forced to relocate to safe houses and commercial areas in the suburbs and the periphery. Some parks remained damaged but intact, but the public squares became no-go zones. Tactics for making sense and making do arose with the immediate need to write new urban space narratives for the city over the ruins of the old, to find occupation in small places, and to provide new centers where the public and the community could meet to return to the familiar aspects of their public lives. Through sampling/recycling/reusing materials at hand, one immediate response was to make small gardens on sites recovered from the rubble, or to leave messages written on walls and any available surface. The contrast between the no-go zones and the places where temporary installations were free for all to traverse was striking. Appropriation and making-do resulted in the speedily constructed Cashel Mall temporary city center, occupying the existing street grid in the adjacent safe zone. Reuse of construction materials at hand, performed in a structured facsimile of the urban street, employed shipping containers, gardens, and enough public space for the community to walk, rest, and recreate. As a symbol of a rebuilt future, the Mall replaced the old center in a completely new form, while the now 'other' city awaited renewal.

This immediate enabling expanded further into the suburbs with the appearance of new insertions into spaces where buildings once were. Sydenham Pod Park, sponsored by local councils and businesses, is a bright urban space that exhibits

Figure 7.5 Christchurch Red Zone border zone, I'll kiss it better, Montage gardens appear where houses once were (2012)

Photo: Gini Lee.

Figure 7.6 Sydenham Pod Park (2015)
Photo: Gini Lee.

some of the dichotomy between new build and the sense of nature taking back the city. Appropriation of readily available materials and ready-made objects repurposed for other uses, provides the structures for a coffee stand, platforms, paving, and pathways through the now greening waste lot; these are tuned in a soft urban landscape that is urban plaza, suburban garden, and community meeting place. The traces of the ruined Beverley Buildings foundations locate the new structures, and it is possible to wander among these and to recall the past life of the site.

Yet the Park is regarded as a transitional place to assist suburban recovery towards a new future, even as it reassembles the past through spatial adaptation, material reuse, and planting regimes and spacing to recall past ecologies of the landscape. For now, the built landscape that was laid waste is recuperated into public space that did not previously exist. The demise of Christchurch's built form has, in the short term, been to the benefit of new forms of public urban space and to a greening of the landscape.

Conclusion: revisiting public space for changing times

One characteristic of uncertain times is revealed in the unstable character of places once conceived for long futures, and predicated upon the spatial and environmental sustainability of the underlying urban landscape. Change scenarios are now invoked by events that supersede original use, resulting in the availability of raw material and/or unconstructed space, readily at hand to enable renewal and reuse of marginal landscapes. Alternative zones of activity arise to enact an interface between the material and ephemeral qualities of the public site and the abstract and concrete conditions that have caused a shift in the territory and its potential uses. Ruined and wasted urban landscapes are post-produced into something else through inventive strategies for activation, informed by techniques that reach far beyond traditional spatial design solutions.

Wastelands are no longer automatically cleared and rebuilt, thus erasing the past. Rather, space and material is appropriated through adopting and creating narratives of occupation and renewal in order to curate events or programs for spaces that endure. The tactics of reuse encourage revisiting places, to view layers through new eyes, and to uncover what is valuable that may have been overlooked. Understanding that all places reveal a palimpsest of occupation is important to uncover and recover the complex systems that are present in such wastelands.

Recycling of materials into new forms and processes suggests both temporary and permanent occupation. The increasingly networked city creates a type of public nomadism demonstrated in the mobile forms that appear as needed and as quickly move on to new sites, thus postproducing an activity in one place for reinvention in another where new forms may result over and over. It is clear that the aesthetics of waste no longer are clearly confined to and defined by spaces regarded as unkempt or marginal. In many ways, such spaces are the new ecological and community zones, reintroduced through educational programming as aligned to subtle and responsive urban and landscape design.

Landscape architects, architects, artists, and cultural geographers, operating in waste theory and practice, all employ on-site locations to amplify their understanding of the notion of waste, beyond the no longer useful object, and into the realm of the public/private landscapes of production and consumption. In many cases, the recovery of sites of abandonment and disuse results in a revised language for that which was once disregarded. We learn from close observation of novel responses to marginalized, disturbed, or overlooked places to adopt methods to postproduce that which already exists in ways that do not pave over the past, but rather enter into an ongoing negotiation that privileges the multiple layers of landscapes of waste over time and space.

Note

1 This is at the lower portion of the map that combines active spaces with aspirations for moments of color and safe passage. From www.capetown.gov.za/en/PublishingImages/ News%20Images/Khayelitsha_upgrade_s.jpg accessed, May 1, 2017.

References

Berger, A. (2006) *Drosscape Wasting Land in Urban America*, New York: Princeton Architectural Press.

Bertagna, A. and Marini, S. (2011) *The Landscape of Waste*, Milan: Skira.

Bocchi, R. (2011) "The Waste Land-scape; Fragment of Thought for a Hypothesis of Landscape as Palimpsest", in A. Bertagna and S. Marini (eds) *The Landscape of Waste*, Milan: Skira.

Bourriaud, N. (2002) *Postproduction: Culture as Screenplay: How Art Reprograms the World*, New York: Lukas and Sternberg.

Calvino, I. (1974) *Invisible Cities*, San Diego: Harcourt Brace and Company.

Clay, G. (1994) *Real Places: An Unconventional Guide to America's Generic Landscape*, Chicago: Chicago University Press.

Engler, M. (2004) *Designing America's Waste Landscapes*, Baltimore: Johns Hopkins University Press.

Fontanari, E (2011) "Urban Arrangements", in A. Bertagna and S. Marini (eds) *The Landscape of Waste*, Milan: Skira.

Gevisser, M. (2014) "My Johannesburg: The City and its Landscape Would Not Exist Were it Not for Many Violations Against Nature", 2 April 2014,www.thenation.com/article/my-johannesburg/, accessed 6 November 2015.

Green, J. (2012) "A New Town for Khayelitsha", https://dirt.asla.org/2012/04/16/a-new-town-for-khayelitsha/A New Town for Khayelitsha04/16/2012Jared Green, accessed 30 January 2016.

Jeffery, C. R. (1977) *Crime Prevention Through Environmental Design*, California: Sage.

Kentridge, W. (2014) *Six Drawing Lessons: The Charles Eliot Norton Lectures*, Boston: Harvard University Press.

Lee, V. (2006) "The Intention to Notice: The Collection, the Tour and Ordinary Landscapes", unpublished PhD thesis, Melbourne: RMIT University.

Lynch, K. (1990) *Wasting Away*, ed. M. Southworth, San Francisco, CA: Sierra Club Books.

Meyer, E. (2008) "Sustaining Beauty. The Performance of Appearance: A Manifesto in Three Parts", *Journal of Landscape Architecture*, vol 3, pp. 6–23.

muf (2010) "Hackney Wick Arts Strategy, Exception Is the Norm: Hackney Wick & Fish Island Design Guidance", September 2010, https://issuu.com/mufarchitectureartllp/docs/designguidance-final_october, accessed 10 May 2015.

muf (2014) "We Are Artists (Pass it on) Since 2010", www.muf.co.uk/portfolio/we-are-artists, accessed 20 July 2016.

Pavement to Parks (2015) *San Francisco Parklet Manual Version 2.2*, http://pavementtoparks.org/wp-content/uploads//2015/12/SF_P2P_Parklet_Manual_2.2_FULL1.pdf, accessed 6 November 2015.

Smithson, R. (1967) "A Tour of the Monuments of Passaic, New Jersey", https://gd1studio2011.files.wordpress.com/2011/09/smithson-monuments-of-passaic.pdf, accessed 20 July 2016.

Umusama (2015) "Violence Prevention Through Urban Upgrading in Khayelitsha, Cape Town, South Africa", https://umusama2015.wordpress.com/2015/04/11/vpuu/, accessed 20 July 2016.

Part 3

Reviving practices of repair and reuse

8 Fix it

Barriers to repair and opportunities for change

Tim Cooper and Giuseppe Salvia

Introduction

Across the industrialized world the amount of repair work undertaken, in relation to the number of goods owned by households, has been in long-term decline (McCollough 2009; Cooper 2010a). This would matter little if such a trend merely reflected an increase in the reliability of goods, whether mechanical, electrical, or electronic, or greater resistance of surface materials to decay or deterioration. If it were so, consumers might be experiencing an increase in product life spans and responding with satisfaction. Yet neither appears to be the case.

Instead, the decline in repair is associated with a throwaway culture (McCollough 2009; Cooper 2013) and shorter product replacement cycles (Bakker et al. 2014). This is contributing to unsustainable consumption in industrialized countries, where the throughput of products – and thus materials – has become faster (Krausmann et al. 2009; Gordon et al. 2006). Moreover, in addition to the environmental damage, many consumers experience frustration because products prove faulty and repair work fails or takes too long (BIS 2014), or is not technically feasible, or would be unreasonably expensive, and is consequently not even attempted.

There is a deeper, systemic malaise underlying this decline: the economic, commercial, and political pressures driving consumerism, which encourage the replacement of products at the earliest opportunity. In the prevailing consumer culture economic performance is judged according to output growth, and the profitability of manufacturers and retailers is dependent upon ever-increasing sales, while politicians pledge to achieve the highest possible 'standard of living', a euphemism for maximum consumption. Hence, consumers have acquired a 'desire for the new' (Campbell 1992), and producers compromise product quality and reparability as they strive to meet ever-tighter 'price points' (Cooper 2012).

This chapter reviews current knowledge on repair activity in the industrialized world, drawing upon a literature base that is currently somewhat limited, reflecting the scarcity of past research. Drawing from multidisciplinary sources, it considers the benefits of repair work in the context of economic, environmental, and social sustainability, identifies barriers to repair activity, and suggests interventions to overcome them. The aim is to explore the complex range of factors underlying repair in order to enable the policy community and other stakeholders to respond to the need to extend product lifetimes, thereby slowing the pace at which goods and materials are consumed (Cooper 2010a). For reasons of brevity, the chapter focuses on household goods (as distinct from vehicles, buildings, infrastructure, industrial goods, and machinery) and literature from the USA and Europe.

The following section defines repair and identifies trends in literature, policy, and practice. In subsequent sections, the potential benefits of repair work are explored and barriers to repair activity identified. Finally, some potential interventions for change are considered.

Understanding repair

Repair is a process whereby a faulty, damaged, or worn product is restored to an acceptable or usable condition and encompasses a complex and fragmented set of activities involving a wide range of organizations and individuals. Commercial repair work is mostly undertaken by (or through) specialist independent repairers, insurance companies (who manage warranties), original equipment manufacturers (or authorized agents), and retailers, sometimes involving third sector (i.e. not for profit) organizations (Parker et al. 2012).

Defining repair and reparability is not always straightforward or objective (Salvia et al. 2015). Repair work includes bonding, sealing, and adjusting or replacing worn or defective parts, but if a material or component that is superior to the original is used, the quality of the product improves and the practice goes beyond restoration and constitutes upgrading. In addition, the distinction between maintenance and repair may be narrow: if maintenance is undertaken too infrequently, the performance or utility provided by a product may cross a threshold beyond which it is considered faulty. The repair process normally, but not necessarily, restores economic value to an item. An ethnographic study by Gregson et al. (2009, pp. 266–267) contrasts restoration "in the manner of a craft consumer" with the "quick-fix mask" of glues and fillers, concluding that repair may be "a means to both object devaluation as well as revaluation."

Likewise, what is deemed 'reparable' may vary according to an owner's values, attitudes, beliefs, habits, attachment to the product, and personal circumstances (e.g. the affordability of a replacement), as well as the prevailing economic and social context (e.g. the cost of repairs, cultural norms). Such thinking is explored by Rosner and Ames (2014) who use the term "negotiated endurance" to describe the process by which the use, maintenance, and repair of products is driven by various actors (consumers, community organizers, and others) through the sociocultural and socioeconomic infrastructures that they inhabit (and produce). Thus the failure of products and their subsequent repair cannot be understood through the lens of design alone.

In a reflective essay Graham and Thrift (2007, p. 1) stress the importance of repair and maintenance to holding off "the constant decay of the world," while Gregson et al. (2009, p. 251) associate maintenance practices with "the conservation and preservation of things." Repair does not merely return products to their past condition, however, but can act as a source of variation, improvisation, and innovation. Maestri and Wakkary (2011) consider home-based design as a creative practice in which objects no longer thought to be useful or usable may be adapted to meet new needs. Citing the argument of Petroski (2006) that 'failure', exemplified by a broken product, should not be regarded as atypical but as a necessary means by which societies, particularly designers, learn, Graham and Thrift (2007, p. 6) conclude:

> Repair and maintenance does not have to mean exact restoration. Think only of the bodged job, which still allows something to continue functioning but probably at a lower level; the upgrade, which allows something to take on new features which keep it contemporary; the cannibalisation and recycling of materials, which allows at least one recombined object to carry on, formed from the bones of its fellows; or the complete rebuild, which allows something to continue in near pristine condition. And what starts out as repair may soon become improvement, innovation, even growth.

Graham and Thrift also note how tinkering with broken consumer goods may not only make them more functional but more personalized, customized, and even redefined. Such thinking was reflected in the work of Netherlands-based design organization Platform21 (2016a), which promoted repairing on the grounds that it could add personality to products and teach users how things are made and how to take care of them, while also giving a sense of accomplishment and control. Similar sentiments have been voiced by another design platform, Mend*RS (2016):

> Mending allows personal material possessions to be recovered, offering them a unique second life. Mended objects bear visible traces of their histories, embody stories of past and present owners, and are imbued with deep emotional value. Mending is an activity which fosters personal knowledge, values, skills and self-efficacy – and thus can be a source of profound satisfaction. Finally, mending provides individuals with the means to foster deeper and longer lasting relationships with the artefacts of their everyday lives.

Nonetheless, a substantial proportion of electrical goods, furniture, and textiles that are discarded as faulty are repairable (Clarke and Bridgwater 2012). In one UK survey, 38 percent of households reported that they "rarely" or "never" had their appliances repaired (Cooper and Mayers 2000), and a more recent UK survey found that nearly 30 percent had disposed of an item because it was broken when, by their own assessment, repair was viable (Pocock et al. 2011). Evidence suggests an unwillingness or inability among consumers to repair small appliances (McCollough 2007) and clothing (WRAP 2012).

When repairs are undertaken, the work can take place *in situ* in the owner's home, on a manufacturer's or retailer's premises, or at a service center, depending on the product type and service system (Parker et al. 2012). Within the EU, most repair work is undertaken by micro enterprises, i.e. those employing fewer than ten people. Survey evidence indicates that the largest repair subsector for "personal and household goods" is electrical household goods, which accounts for around one half of the total "value added" and employment; other specified subsectors, such as footwear, clocks, watches, and piano tuning, are considerably smaller (Eurostat 2016). Electrical and electronic items are thus given particular attention in this chapter.

The benefits of repair

The repair of faulty or damaged goods has the potential to contribute to economic, environmental, and social sustainability. Reflected in the European Commission's *Action Plan for the Circular Economy*, this has led to a growing international debate:

> Once a product has been purchased, its lifetime can be extended through reuse and repair, hence avoiding wastage. The reuse and repair sectors are labour-intensive and therefore contribute to the EU's jobs and social agenda.
> (European Commission 2015, p. 7)

Repair and maintenance activity remains important to the industrialized economies, particularly with regard to employment, despite its long-term decline in many product sectors (Bakker et al. 2014; Cooper 2006). In the USA, around 1.3 million people are employed in various repair and maintenance sectors: commercial and industrial machinery, electronic and precision equipment, automotive, and personal and household goods (Bureau of Labor Statistics 2016). Likewise, in the European Union around 1.5 million people are employed in the maintenance and repair of motor vehicles and nearly 400,000 in the repair of computers and personal and household goods (Eurostat 2016). Moreover, the EU data only refers to work undertaken professionally by specialist repairers; it excludes enterprises undertaking repair as a secondary activity and repair work by owners.

Repair also offers potential environmental benefits in the form of lower materials consumption and, ultimately, less waste; it consequently forms an important element in strategies to achieve resource efficiency and a circular economy (European Commission 2012, 2015). Reduced materials consumption is important in the context of resource security, and because of the energy used when extracting and refining materials and during manufacturing and distribution (which is thus 'embodied' in products). Addressing embodied carbon by tackling consumption is increasingly recognized as necessary in the climate change debate (Cranston and Hammond 2012). For instance, the repair, refurbishment, or remanufacture of computer hardware is estimated to generate less than 20 percent of the CO_2 emissions emitted through manufacturing an equivalent new product (BSI 2012). Furthermore, in the context of closed-loop strategies, repair is generally considered preferable to reconditioning or remanufacturing because less energy is needed and virtually all material is kept in use (King et al. 2006). Even so, environmental impacts will depend on how and where repair work is undertaken and cannot be assumed to be beneficial. For example, there may not be a net reduction in energy consumption if service engineers travel to mend products *in situ*, and appliances are substantially less energy efficient than potential replacement models, or a repaired products proves short-lived.

By reducing materials consumption, repair should reduce the volume of waste (although, again, if savings in cost are spent on new products, this cannot be assumed). Thus, in the waste management hierarchy that is commonly applied to waste policy, as in Article 4 of the EU Directive on Waste (2008/98/EC) (as the Waste Framework Directive), repair is implicit in the highest category (prevention), explicit in the second (preparing for reuse), and ahead of recycling, energy recovery, and disposal. Article 11 requires Member States to encourage the establishment and support of reuse and repair networks. Article 29 requires them to establish waste prevention programs and proposes the promotion of reuse and repair "through the use of educational, economic, logistic or other measures, such as support for the establishment of accredited repair and reuse centers and networks, especially in densely populated regions" (paragraph 16, Annex IV).

Lastly, repair has potential in the less well-defined areas of social sustainability (McKenzie 2004) to offer benefits in connection with people's quality of life (their sense of well-being and material security), equity (issues of fairness and affordability), social cohesion (through community or other networks), and participation (collaborative engagement with products).

Thus the process of repair is not merely a functional phenomenon but is significant in terms of people's well-being, particularly when they have a strong emotional attachment to a product (Mugge et al. 2005). Repair can also have implications for equity, in that the supply and cost of repair services is especially important to people unable to afford new products. As individual's incomes increases, they become more likely to replace products than repair them (McCollough 2007).

Repair also offers benefits in terms of social cohesion. Repair work is often a financially marginal activity and in recent years has increasingly been undertaken by social enterprises that train disadvantaged individuals and sell (or give) the repaired items to people in low income households, many of those in the UK linked to the Furniture Reuse Network (FRN 2015). Lastly, community benefits can be seen in the grassroots social innovation of Repair Cafés and similar initiatives. Such initiatives are essentially participatory, enabling people to engage directly with their broken products and mend them with the support of volunteers who provide the necessary repair rather than passing them on to commercial 'experts' (Rosner 2012; Charter and Keiller 2014).

Responses to the downward trend

Despite these benefits, repair activity appears to be in long-term decline. In the UK, household expenditure on repair work fell more than 40 percent over a 40-year period, from 1964 to 2004 (Cooper 2006). The decline was observed across various product categories; it was particularly marked for footwear and, while fluctuating somewhat, was also evident for household appliances and audio-visual, photographic, and information processing equipment. A similar downward trend has been seen in the USA, where the number of service centers for appliances and electronic goods more than halved over the 15-year period up to 2007 (McCollough 2010). Such trends are all the more significant when viewed in the context of increased consumption.

Downward trends may, to some degree, reflect improvements in product reliability. Which?, the UK-based consumer organization, has produced reports on product reliability over many years. In a recent review, it concluded that since 1971 "while there have been ups and downs, appliances have generally become more reliable" (Which? 2011, p. 20). Its comparison of survey results from 2001 to 2011 confirmed the trend: in 2001 21 percent of washing machines aged up to six years old had to be repaired, but by 2011 the figure had dropped to 12 percent; reductions were also recorded for dishwashers (17 percent to 11 percent), tumble dryers (12 percent to 10 percent) and washer dryers (29 percent to 21 percent). However, the review also noted a change in consumer attitudes and behavior:

> We are now far more ready to throw away products, rather than repair them . . . Consumer attitudes to repairing or replacing domestic appliances are changing – fewer are repaired when they break down, replacing broken machines is on the rise and twice as many appliances are being replaced when they're still working, compared with the 1950s.
>
> (Which? 2011, pp. 20–21)

Few attempts have been made to counter this trend and not all have survived. One of the most significant was the aforementioned Platform21, a short-lived design venture in the Netherlands. Exhibitions, research projects, workshops, lectures, discussions, and club nights were organized. At the latter, participants sought to "question today's society, connect amateur and professional creativity, reveal the making process, and stimulate dialogue and the sharing of creative knowledge" (Platform21 2016b). A key output was its *Repair Manifesto*, which urged designers and consumers "to break the chain of throwaway thinking" and was downloaded more than one million times (Platform21 2016a).

Perhaps most influential is the growing network of Repair Cafés, a concept based around a space where people meet to socialize and repair various consumer goods such as clothes, furniture, electrical appliances, bicycles, crockery, appliances, and toys. Having originated in 2009 in the Netherlands, by 2016 there were over 1,000 Repair Cafés operating in 24 countries (Repair Café 2016).

Another solution for reversing the decline in repair has been the development of home-based repair supported by networked learning, through which owners of faulty or damaged goods gain the knowledge and skills necessary to mend them from dedicated websites. The highest profile organization in this arena is iFixit, a Californian company formed in 2003 that publishes free online repair manuals for consumer electronics and sells repair parts on its wiki-based web site. iFixit has more than one million members, the majority outside of the USA (Sabbaghi et al. 2016), and to date has produced nearly 20,000 repair manuals (iFixit 2016).

Such developments have attracted the attention of some major businesses. At a recent conference, Ikea's Head of Sustainability argued that societies are not only confronted with "peak oil" but "peak stuff" and revealed that the company "will be increasingly building a circular Ikea where you can repair and recycle products" (Farrell 2016). A growth in web-based initiatives to provide after-sales advice and services aimed at independent repairers (e.g. White Goods Help 2016) has become apparent.

There has also been a response from government bodies and campaigners. In 2013 the UK Government was the first EU Member State to produce a waste prevention program as required by the Waste Framework Directive (2008/98/EC) (HM Government 2013). Titled *Prevention Is Better Than Cure*, it presents various measures to promote repair: the development of a web-based tool to enable householders to find local reuse and repair services, plans to bring various stakeholders together to facilitate greater reuse and repair, working with industry to develop more services and standards for repair and reuse, and research aimed at a better understanding of the repair sector. According to Lane and Watson (2012), however, significant research on recycling and reuse has already been conducted usign a range of analytical scales – the macro-scale of changing waste management regimes, the meso-scale of the household and neighborhood, and the micro-scale of individual behavior – but the findings have not yet been applied to policy.

Following a recommendation by the European Economic and Social Committee (2013) that companies should make their products easier to repair, the European Commission's *Action Plan for the Circular Economy* included a testing program to identify issues concerning planned obsolescence and proposed future work on ecodesign:

> Currently, certain products cannot be repaired because of their design, or because spare parts or repair information are not available. Future work on ecodesign of products will help to make products more durable and easier to repair: in particular, requirements concerning the availability of spare parts and repair information (e.g. through online repair manuals) will be considered, including through exploring the possibility of horizontal requirements on the provision of repair information.
>
> (European Commission 2015, p. 7)

The proposed work would be based on the Ecodesign Directive (2009/125/EC), which empowers the EU to issue requirements for specified energy-related products placed on the internal market concerning the availability of spare parts, modularity, upgradeability, and reparability. In particular, demands from independent repairers for greater access to product information have grown in recent years. For example, following a sustained "right to repair" campaign, car manufacturers in Massachusetts, USA are now required to provide the same information to independent repairers as they offer to their dealers and authorized repair facilities (RREUSE 2015).

While the European Commission's *Action Plan* represented a degree of progress, it did not take forward many of the proposals to use existing European legislation to improve the reparability of products suggested in a report by RREUSE (2015), an umbrella organization representing social enterprises active in reuse, repair, and recycling. To date, few of the many policy options that could encourage repair (Cooper 2010b) have been translated into legislation. One exception is an EU Directive (1999/85/EC) which enables reduced rates of VAT to be levied on labor-intensive services including "minor repairing" of bicycles, shoes and leather goods, and clothing and household linen. Initially introduced for a limited period and evidently motivated by a desire to test the potential for job creation and to combat the black economy rather than any environmental benefit, the arrangement was made permanent by a subsequent EU Directive (2009/47/EC). Currently, however, only eight Member States make use of this opportunity. Another policy approach is the inclusion of reparability among criteria for products to be eligible for an EU Ecolabel (Regulation (EC) No 66/2010). Thus, in the case of televisions and personal computers, for example, a manufacturer must demonstrate that the item can be easily disassembled, and spare parts must be available for a specified period after production has ceased (seven years for televisions, five years for personal computers).

Barriers to repair work

The level of repair work undertaken in industrialized economies depends on a range of interconnected variables. Some relate to the product in question, specifically its reliability, resistance to 'wear and tear' and whether the design has allowed for ease of repair. Others relate to the owners' propensity to repair faulty or damaged items, which will depend on their values, attitudes, skills, and knowledge, and may be influenced by factors such as cost and convenience. Other variables relate to the wider context, such as the supply of repair services by manufacturers, retailers, and independent repairers and socio-cultural norms and expectations.

In this section, the barriers to repair work are considered in relation to these variables, while the following section addresses how the barriers might be overcome, with reference to government support, business models, and grassroots developments.

The product: inappropriate design

Design is critical to reparability – the potential for disassembly, use of irreversible closures, multiple or low quality materials, non-standardized parts, surface coatings, glues and welding, together with insufficient information on materials and inadequate labeling, may affect the ease with which repair can be undertaken (Cooper 2006; Vezzoli and Manzini 2008; Parker et al. 2012).

Critics sometimes portray design that precludes or hinders repair as a deliberate strategy of planned obsolescence (Packard 1961; Slade 2006). Deliberate or not, such design is a logical outcome of a mass production system in which standards are sometimes compromised to meet the 'price point' necessary to attract consumers (Cooper 2012). Even when designers want products to be easily repaired, they may face problems due to technology limitations and safety concerns (Cohen-Rosenthal 2004). In the electronics sector, in particular, products have become more sophisticated and often miniaturized, often evolving faster than the ability of users to learn how to repair them.

A survey of UK repairers (Which? 2011) supports the common assumption that household appliances have become harder to repair, while a report by RREUSE (2013) has detailed obstacles to repairing fridges, dishwashers, and washing machines. These include rapid changes in product design and components, lack of access to reasonably priced spare parts, lack of access to essential repair and service manuals, software and hardware, and difficulty in separating individual parts from the casing or accessing key internal components.

The owner: propensity and ability to repair

Personal traits affect people's propensity to repair goods (Scott and Weaver 2014). Some people associate restoring products to a sound condition with a past era of 'make do and mend' and, in contrast with contemporary wastefulness, responsible citizenship. They may also see repair as a form of craft production (Gregson et al. 2009) that makes good use of skilled manual labor and, when undertaken satisfactorily, creates satisfaction for the repairer, whether professional or householder.

On the other hand, experiences of professional repair work proving more expensive than anticipated, or a product not restored to an acceptable condition, can undermine people's trust in the repair process and reduce their willingness to consider it in future (Defra 2011). The propensity of consumers to have broken or damaged products repaired may be also linked to factors such as income: a Scottish survey found that people in lower socio-economic groups are less likely to repair household items if they break, and that other socio-demographic variables, such as gender, influence whether particular types of item are repaired (Lee-Woolf et al. 2012).

Uncertainty and risk will inhibit some people from getting products repaired. They may feel vulnerable because of asymmetry in repair knowledge (i.e. professionals generally know more about the product than them), particularly if it is not possible to discern the quality of the repair work immediately (McCollough 2010). If charges for labor and parts are not identified separately, as is common, they may be uncertain as to whether the charge is fair. They may fear that replacement components are substandard (or, in the case of authorized parts, overpriced).

Consumers will be more likely to sense a degree of risk and be reticent to have products repaired if they do not know a local independent repairer with a good reputation. Increased trust in the service supplier reduces the level of perceived risk (Shemwell et al. 1994). Consumer protection legislation provides only partial reassurance. A report from 1988 on the servicing of cars and electrical appliances concluded that consumers were ill-informed about the possibility and conditions of after-sales services: tariffs were rarely published, estimates were not obligatory, there was uncertainty about the implications of unsuccessful repair work, and the period within which it should be completed was unspecified (European Consumer Law Group 1988). Twigg-Flesner (2010, p. 212) has argued that legislation in the UK addresses "only some of the problems highlighted two decades ago," describing consumer protection under the *Supply of Goods and Services Act 1982* as "rather basic." For example, while a person carrying out repair work is required to provide the service "with reasonable care and skill" (Section 13), there remains potential for uncertainty when a repair is attempted but does not prove effective.

Deciding whether to have a product repaired can also be problematic due to a lack of information. The cost may be unclear if the diagnosis is uncertain, while cost-effectiveness will be hard to predict when the residual life span is unknown. Many people are unsure about typical product life spans, and expectations vary (Defra 2011; Cooper and Christer 2010). Consumers also apply different subjective discount rates to their possessions, depending on how they value them (McCollough 2010).

Gregson et al. (2009) document routinized practices of repair and maintenance of household objects and describe how people's skills (such as carpentry or sewing) and the physical spaces of their dwellings facilitate these activities. While home repairs may still be commonplace (Lee-Woolf 2012), many people lack the knowledge and skills necessary to maintain products in good working order. The offshoring of manufacturing has led to a decline in manual skills that, in the past, were transferred to a domestic context. The reduction in design and technology education in the UK reinforces this problem, leaving young people less confident in their ability to repair goods. In addition, manufacturers of electrical goods often dissuade users from repairing products, for example, making warranty provision void if repairs are attempted.

The context: repair services, economics, and socio-cultural norms and expectations

In stark contrast to the ease with which new products can be purchased, finding a repairer who is well qualified, reliable, and available may be hard, and the repair process itself may involve lengthy travel and waiting times. Graham and Thrift (2007) noted an uneven distribution of reuse and repair centers, while Scott and Weaver (2014) concluded that repair is often time-consuming and inconvenient. Local repair services suffer from inadequate support and training: many repairers are independent traders and lack management skills and marketing capacity (Lee-Woolf et al. 2012), and some depend on receiving information or spare parts from manufacturers.

The economics of repair are crucial. Repair work can be expensive because of high labor costs, design that makes disassembly time-consuming, use of expensive sealed units, and over-priced spare parts (Parker et al. 2012). In addition, payroll taxes associated with employment (which are typically used to fund social security systems) discourage labor-intensive work such as repair. That said, there is evidence that consumers perceive the cost of repair to be higher than it is in reality (Huysentruyt and Read 2010).

Replacement is increasingly preferred to repair when faults arise because many new goods have become progressively cheaper (Cooper 1994, 2010b; Downes et al. 2011; Lee-Woolf 2012). Consumers are only willing to pay a small fraction of the price of a replacement good for repair work, around 20 percent in the USA in the case of appliances (McCollough 2009). In practice, many such decisions are based on hunches rather than firm data – consumers do not have access to the 'whole life costing' information necessary to make economically 'rational' choices. Thus, some apply a rule of thumb that a broken product is beyond economic repair when the cost would exceed the resale value of the item. A lack of market transparency is another problem; prices vary significantly between manufacturers and independent repairers, and also among independent repairers (OFT 2011). Extended warranties reduce the risk of high repair costs and spread payments over time, but are generally criticized by consumer organizations as representing poor value (Which? 2013).

Lastly, socio-cultural norms and expectations can constitute a further barrier. In a consumer culture in which many people feel judged by what they own and their ability to purchase the latest products, the pressure to purchase a new item if a product develops a fault, rather than have it repaired, may be substantial (Defra 2011).

Opportunities for change

Reviving a repair culture that has been in decline for over 50 years clearly represents a major challenge. Although signs of change are emerging, a combination of governmental, business and grassroots, community-led interventions appear necessary if the momentum is to increase. In this final section, some potential initiatives are proposed for these three actors.

Increased government support

The role of repair in enabling a more sustainable society has begun to receive unprecedented attention from governments through the aforementioned waste prevention programs required of EU Member States and the European Commission's *Action Plan for the Circular Economy*. Even so, much more could be done to promote product reparability according to a series of reports published prior to the launch of the Action Plan by RREUSE, the European Environmental Bureau and the BEUC (Bureau Européen des Unions de Consommateurs).

The report by RREUSE (2015), which represents 29 national and regional networks active in reuse, repair, and recycling, with around 77,000 employees and over 60,000 volunteers and trainees, suggested making use of existing European legislation to promote product reparability. It proposed a "right to repair" approach to enable independent service engineers to have access to the repair and diagnostic information for products provided by manufacturers to their dealers and authorized repair facilities, applying Article 15 of the WEEE Directive in the case of electrical and electronic goods. Noting criteria used in the Austrian "Label of excellence for durable, repair-friendly designed electrical and electronic appliances" (ONR 192102), the RREUSE Report suggested utilizing provisions in the Ecodesign Directive (2009/125/EC) to require ease of disassembly and use of standardized components where appropriate (e.g. screws, motors, and pumps). In addition, it proposed using the Consumer Rights Directive (2011/83/EC) to make sellers liable for defects for longer than the current two-year period and broadening the criteria currently used in the Energy Labelling Directive (2010/30) to require information on durability and reparability.

The European Environmental Bureau (EEB), a federation of over 140 environmental citizens' organizations, similarly published a package of measures aimed at increased product durability and reparability (EEB 2015). Its report recommended a system to rate the durability and reparability of products (with associated standards for measurement), longer minimum legal warranty periods (three to ten years, depending on product category), a public communication campaign to highlight the opportunities and benefits of reuse and repair, and various design and information requirements. Meanwhile, BEUC, which represents European consumer organizations, urged "action for a new culture of reparability" (BEUC 2015, p. 14) and outlined a range of policy measures relating to the Ecodesign Directive, consumer information on product lifetimes, legal guarantees, and the availability of spare parts and software support.

As the cost of repair work, relative to replacement, is an important deterrent to consumers (Scott and Weaver 2014), governments need to develop economic policies to incentivize repair. As noted above, the European Union has taken tentative steps in this direction with the aforementioned legislation enabling Member States to reduce VAT on repairs to certain products. Further evidence may be needed to persuade more Member States to participate, and the overall impact would be greater if a lower rate could be applied to other goods, such as computers and watches. Indeed, RREUSE (2015) proposed that there should be zero VAT on all repair, maintenance, and upgrade services. Another measure to promote such work would be to reduce taxation on labor which, according to research by the European Commission (2003), may prove more effective than reducing VAT.

Lastly, governments could set an example through their procurement policies, purchasing or hiring products designed for longevity and reparability, and prioritizing repair over replacement where this is a credible option.

Innovative business models

Businesses will only move from using a conventional sales-driven model to one in which a greater share of revenue is generated from product life extension activities such as repair and maintenance if they foresee commercial benefits. This process, sometimes described as 'servitisation', has long been discussed but is still rare in consumer goods markets. One approach might be to encourage companies to consider repair as a means to build brand loyalty (Scott and Weaver 2014; Sabbaghi et al. 2016).

WRAP (Waste and Resources Action Programme), a UK charity that works in partnership with businesses to promote a more resource-efficient economy, has identified several business models that could increase repair opportunities for different types of appliance (WRAP 2016). One proposal is that existing reuse organizations expand their activities to include the repair of white goods; this would increase their revenue while requiring minimum set-up costs. In a second, television repair services are offered at a fixed price, based on screen size, in order to overcome the barrier of price uncertainty, and costs are reduced by repairing faulty components rather than replacing them. In a third proposal, faulty tablet computers are swapped for an equivalent model, enabling customers to have a replacement tablet quicker and at a lower cost than repairing the original; faulty models are then repaired in bulk for future distribution.

Stronger relationships between manufacturers, retailers and social innovators may help to accelerate progress towards new business models. For example, the aforementioned UK-based social enterprise network FRN (Furniture Reuse Network) works in partnership with companies such as Ikea, Dixons Carphone, and John Lewis to collect unwanted products for subsequent reuse, repairing them first when necessary (FRN 2015).

Local government, too, has the potential to play an important role. In Austria, municipal authorities in the three largest cities (Vienna, Linz, Graz), working together with several regional administrations, have developed a guide to repair, lending and second-hand goods that contains repair tips and the addresses of relevant companies. One such enterprise, R.U.S.Z. (*Reparatur und Service Zentrum*) – a repair and service center which trains people without jobs to repair household appliances and consumer electronics – helped to initiate the Repair Network Vienna, a network of some 60 private, profit-oriented repair companies which is reported to have substantially increased demand for repair services (Pre-waste 2010).

Grassroots developments

Finally, the future level of repair activity will partly depend on the success, or otherwise, of grassroots, often community-led, initiatives. These include Repair Cafés, Fixit Clinics, and similar local ventures, and user repair enabled by networked learning.

Repair Cafés have grown significantly in number, as described above, and may be seen as part of an evolving shift in people's relationships towards (and interactions with) products, which is also reflected in developments such as Maker Fairs, Hackerspaces, and vintage retailing. This appears to be a values-driven cultural trend (Lee-Woolf 2012). In a study of community repair in the USA, Rosner and

Turner (2015, p. 62) identified "a surprising connection between repair work and social movements associated with environmentalism and sustainability . . . Participants believe that their acts of repair constitute interventions in large-scale social processes and that they can have effects far beyond their local setting." An international study found that Repair Café volunteers and organizers are "strongly motivated to take part because of what they can do for others, namely their desire to help others live more sustainably, to provide a valuable service to the community and to help improve product reparability and longevity" (Charter and Keiller 2014, p. 2). Asked about their expectations for the next five years, respondents anticipated linking Repair Cafés to form more effective local repair networks and greater involvement with campaigning to improve product reparability and longevity. There was also an expectation that Repair Cafés would prompt business start-ups and develop closer links with commercial repairers.

Whether this phenomenon of community-led repair initiatives will expand sufficiently to have more than a marginal impact on overall consumption remains uncertain. Seyfang and Smith (2007) concluded that emerging grassroots movements sometimes appear promising but are liable to have a slow uptake without regulatory frameworks that support their work, new economic models, and major changes in consumer behavior.

Although the extent of home-based repair work is commonly assumed to have declined, evidence of change is emerging. The growth of networked learning, through which people learn to repair products using websites such as that of iFixit (2016), could become highly significant. Scaling-up repair by users would require better access to product information and reasonably priced spare parts (RREUSE 2013). Such change will also demand more positive consumer attitudes towards repair – McCollough (2009) proposed an expansion of vocational and home economics courses in schools to improve people's skills.

Conclusion

The potential contribution of repair work to economic, environmental, and social sustainability is increasingly accepted. A trend from replacement to repair when products develop faults could, in principle, increase employment, reduce materials consumption and waste, and provide social benefits.

This chapter has explained the historic long-term decline in repair activity and current barriers to change with reference to the economic, psychological, socio-cultural, and political forces that continually drive sales of new, replacement products in a culture of mass production, and unsustainable consumerism. Changing this culture and its underlying multi-faceted obsolescence will require radical interventions. Governments and businesses have begun to engage with this agenda but have yet to implement many of the necessary measures. Meanwhile, a deeper relationship between people and products is also required, a greater emotional attachment that leads owners to maintain products for longer and, if they break, repair them. Again, there are signs of change, particularly in the development of grassroots initiatives, but these have yet to extend beyond the fringes of society.

Promising interventions that could transform the dominant throwaway culture, scale-up user demand for repair, and improve repair services have been identified. They include tax reform to create better economic incentives, design for ease of disassembly and greater standardization of commonly used parts, wider access to product information and diagnostic tools, longer guarantees, innovative business models involving closer relationships with social enterprises, and improved design-related education. Whether such interventions ultimately succeed will be determined by societal attitudes towards consumption and, in particular, people's relationships with their possessions – and the willingness of companies to change their business models.

Acknowledgment

The research for this chapter was undertaken with financial support from the EPSRC, grant reference EP/N022645/1.

References

Bakker, C., Wang, F., Huisman, J. and den Hollander, M. (2014) "Products That Go Round: Exploring Product Life Extension Through Sesign", *Journal of Cleaner Production*, vol 69, pp. 10–16.
BEUC (Bureau Européen des Unions de Consommateurs) (2015) *Durable Goods: More Sustainable Products, Better Consumer Rights*, BEUC-X-2015–069, www.beuc.eu/publications/beuc-x-2015-069_sma_upa_beuc_position_paper_durable_goods_and_better_legal_guarantees.pdf, accessed 28 October 2016.
BIS (Department for Business, Innovation and Skills) (2014) *Consumer Engagement and Detriment Survey 2014*, London: Department for Business, Innovation and Skills.
Brook Lyndhurst (2011) *Public Understanding of Product Lifetimes and Durability*, Report by Brook Lyndhurst for Defra, www.brooklyndhurst.co.uk/public-understanding-of-product-lifetimes-and-durability-_156.html, accessed 28 October 2016.
BSI (British Standards Institution) (2012) *BS 8887–211:2012 Design for Manufacture, Assembly, Disassembly and End-of-Life Processing (MADE) Specification for Reworking and Remarketing of Computing Hardware*, London: BSI.
Bureau of Labor Statistics (2016) *Industries at a Glance – Repair and Maintenance: NAICS 811*, www.bls.gov/iag/tgs/iag811.htm, accessed 28 October 2016.
Campbell, C. (1992) "The Desire for the New", in R. Silverstone and E. Hirsch (eds) *Consuming Technologies*, London: Routledge.
Charter, M. and Keiller, S. (2014) *Grassroots Innovation and the Circular Economy: A Global Survey of Repair Cafés and Hackerspaces*, July 2014, Farnham: The Centre for Sustainable Design/University for the Creative Arts, www.research.ucreative.ac.uk/2722/1/Survey-of-Repair-Cafes-and-Hackerspaces.pdf, accessed 28 October 2016.
Clarke, E. and Bridgewater, E. (2012) *Composition of Kerbside and HWRC Bulky Waste*, Report by Resource Futures for WRAP, Banbury, UK: WRAP.
Cohen-Rosenthal, E. (2004) "Making Sense Out of Industrial Ecology: A Framework for Analysis and Action", *Journal of Cleaner Production*, vol 12, no 8–10, pp. 1111–1123.
Cooper, T. (1994) "The Durability of Consumer Durables", *Business Strategy and the Environment*, vol 3, no 1, pp. 23–30.

Cooper, T. (2006) *Repair Activity in the UK*, Working Paper 1 for the ESRC Centre for Business Relationships, Accountability, Sustainability and Society (BRASS), Sheffield: Sheffield Hallam University.

Cooper, T. (2010a) "The Significance of Product Longevity", in T. Cooper (ed) *Longer Lasting Products: Alternatives to the Throwaway Society*, Farnham: Gower.

Cooper, T. (2010b) "Policies for Longevity", in T. Cooper (ed) *Longer Lasting Products: Alternatives to the Throwaway Society*, Farnham: Gower.

Cooper, T. (2012) "The Value of Longevity: Product Quality and Sustainable Consumption", *Proceedings: Global Research Forum on Sustainable Consumption and Production*, 13–15 June, Rio de Janiero, Brazil, http://grf-spc.weebly.com/uploads/2/1/3/3/21333498/_cooper-paper.pdf, accessed 28 October 2016.

Cooper, T. (2013) "Sustainability, Consumption and the Throwaway Culture", in S. Walker and J. Giard (eds) *The Handbook of Design for Sustainability*, London: Bloomsbury.

Cooper, T. and Christer, K. (2010) "Marketing Durability", in T. Cooper (ed) *Longer Lasting Products: Alternatives to the Throwaway Society*, Farnham: Gower.

Cooper, T. and Mayers, K. (2000) *Prospects for Household Appliances*, Halifax: Urban Mines, http://irep.ntu.ac.uk/6671/, accessed 28 October 2016.

Cranston, G. R. and Hammond, G. P. (2012) "Carbon Footprints in a Bipolar, Climate-constrained World", *Ecological Indicators*, vol 16, pp. 91–99.

Downes, J., et al. (2011) *Longer Product Lifetimes, Chapter 1 – Scoping Exercise*, February, Report by ERM for Defra, http://sciencesearch.defra.gov.uk/Default.aspx?Menu=Menu&Module=More&Location=None&Completed=0&ProjectID=17047, accessed 28 October 2016.

EEB (European Environmental Bureau) (2015) *Circular Economy Package 2.0: Some Ideas to Complete the Circle*, www.eeb.org/index.cfm?LinkServID=1E2E1B48-5056-B741-DB594FD34CE970E9, accessed 28 October 2016.

European Commission (2003) *Report from the Commission to the Council and the European Parliament – Experimental Application of a Reduced Rate of VAT to Certain Labour-intensive Services* [SEC(2003) 622], COM/2003/0309 final.

European Commission (2012) *Roadmap to a Resource Efficient Europe*, Communication from the Commission to the European Parliament, the Council, the European Economic and Social Committee and the Committee of the Regions, COM/2011/0571 final.

European Commission (2015) *Closing the Loop: An EU Action Plan for the Circular Economy*, Communication from the Commission to the European Parliament, the Council, the European Economic and Social Committee and the Committee of the Regions, COM(2015) 614 final.

European Consumer Law Group (1988) "Servicing of Cars and Electrical Appliances", in European Consumer Law Group (ed) (1997) *Reports and Opinions 1986–97*, Louvain-la-Neuve: Centre de Droit de la Consommation, pp. 105–136.

European Economic and Social Committee (2013) *Towards More Sustainable Consumption: Industrial Product Lifetimes and Restoring Trust Through Consumer Information*, Opinion of the European Economic and Social Committee and the Committee, CCMI/112, www.eesc.europa.eu/?i=portal.en.ccmi-opinions.26788, accessed 28 October 2016.

Eurostat (2016) "Statistics Explained", http://ec.europa.eu/eurostat/statistics-explained/index.php/Statistics_Explained, accessed 28 October 2016.

Farrell, S. (2016) "We've Hit Peak Home Furnishings, Says Ikea Boss", *The Guardian*, 18 January 2016, www.theguardian.com/business/2016/jan/18/weve-hit-peak-home-furnishings-says-ikea-boss-consumerism, accessed 28 October 2016.

FRN (Furniture Reuse Network) (2015) "Commercial Retailers: Their Impact on the UK Reuse Sector", www.frn.org.uk/images/frn/FRN%20Commercial%20Impact%20 Report%202015%20web.pdf, accessed 28 October 2016.

Gordon, R. B., Bertram, M. and Graedel, T. E. (2006) "Metal Stocks and Sustainability", *Proceedings of the National Academy of Sciences of the USA*, vol 103, no 5, pp. 1209–1214.

Graham, S. and Thrift, N. (2007) "Out of Order: Understanding Repair and Maintenance Theory", *Culture and Society*, vol 24, no 3, pp. 1–25.

Gregson, N., Metcalfe, A. and Crewe, L. (2009) "Practices of Object Maintenance and Repair: How Consumers Attend to Consumer Objects Within the Home", *Journal of Consumer Culture*, vol 9, no 2, pp. 248–272.

HM Government (2013) *Prevention Is Better Than Cure: The Role of Waste Prevention in Moving to a More Resource Efficient Economy*, www.gov.uk/government/uploads/ system/uploads/attachment_data/file/265022/pb14091-waste-prevention-20131211.pdf, accessed 28 October 2016.

Huysentruyt, M. and Read, D. (2010) "How Do People Value Extended Warranties? Evidence from Two Field Surveys", *Journal of Risk and Uncertainty*, vol 40, pp. 197–218.

iFixit (2016) "iFixit", www.ifixit.com/, accessed 28 October 2016.

King, A. M., Burgess, S. C., Ijomah, W. and McMahon, C. A. (2006) "Reducing Waste: Repair, Recondition, Remanufacture or Recycle?", *Sustainable Development*, vol 14, no 4, pp. 257–267.

Krausmann, F., Gingrich, S., Eisenmenger, N., Erb, K. H., Haberl, H. and Fisher-Kowalski, M. (2009) "Growth in Global Materials Use, GDP and Population During the 20th Century", *Ecological Economics*, vol 68, no 10, pp. 2696–2705.

Lane, R. and Watson, M. (2012) "Stewardship of Things: The Radical Potential of Product Stewardship for Re-framing Responsibilities and Relationships to Products and Materials", *Geoforum*, vol 43, no 6, pp. 1254–1265.

Lee-Woolf, C., C., Hughes, O., Fernandez, M. and Cox, J. (2012) *Engagement with Re-use and Repair Services in the Context of Local Provision: Exploring the Relationship Between Re-use and Repair Behaviours and the Provision of Services in Different Areas Across Scotland*, Report by Brook Lyndhurst for Zero Waste Scotland, www.brooklynd hurst.co.uk/re-use-and-repair-behaviour-in-context-_201, accessed 28 October 2016.

Maestri, L. and Wakkary, R. (2011) "Understanding Repair as a Creative Process of Everyday Design", *Proceedings of the 8th ACM Conference on Creativity and Cognition*, Association for Computing Machinery, 3–6 November 2011, Atlanta, Georgia, USA.

McCollough, J. (2007) "The Effect of Income Growth on the Mix of Purchases Between Disposable Goods and Reusable Goods", *International Journal of Consumer Studies*, vol 31, no 3, pp. 213–219.

McCollough, J. (2009) "Factors Impacting the Demand for Repair Services of Household Products: The Disappearing Repair Trades and the Throwaway Society", *International Journal of Consumer Studies*, vol 33, no 6, pp. 619–626.

McCollough, J. (2010) "Consumer Discount Rates and the Decision to Repair or Replace a Durable Product: A Sustainable Consumption Issue", *Journal of Economic Issues*, vol 44, no 1, pp. 183–204.

McKenzie, S. (2004) *Social Sustainability: Towards Some Definitions*, Hawke Research Institute Working Paper Series No 27, www.sapo.org.au/pub/pub241.html, accessed 28 October 2016.

Mend*RS (2016) Inaugural Mend*RS Workshop, http://discardstudies.com/2012/03/21/ inaugural-mendrs-workshop/, accessed 28 October 2016.

Mugge, R., Schoormans, J. P. L. and Schifferstein, H. N. J. (2005) "Design Strategies to Postpone Consumers' Product Replacement: The Value of a Strong Person-Product Relationship", *The Design Journal*, vol 8, no 2, pp. 38–48.

OFT (Office of Fair Trading) (2011) *Aftermarkets for Domestic Electrical Goods*, OFT1320, April 2011, http://webarchive.nationalarchives.gov.uk/20140402142426/ http:/oft.gov.uk/OFTwork/markets-work/othermarketswork/electrical-goods/#named7, accessed 28 October 2016.

Packard, V. (1961) *The Waste Makers*, London: Longmans.

Parker, D., et al. (2012) *Understanding the Opportunities to Increase Re-use and Repair*. Report by Oakdene Hollins, Brook Lyndhurst and Nottingham Trent University for WRAP, Banbury, UK: WRAP.

Petroski, H. (2006) *Success Through Failure*, Princeton: Princeton University Press.

Platform21 (2016a) "Platform21 = Repairing", www.platform21.nl/page/4315/en, accessed 28 October 2016.

Platform21 (2016b) "About Us", www.platform21.nl/page/133/en, accessed 28 October 2016.

Pocock, R., Clive, H., Coss, D. and Wells, P. (2011) *Realising the Reuse Value of Household WEEE*, Report by M·E·L Research for WRAP, www.wrap.org.uk/sites/files/wrap/WRAP%20WEEE%20HWRC%20summary%20report.pdf, accessed 28 October 2016.

Pre-waste (2010) "Mapping Report on Waste Prevention Practices in Territories Within EU27", www.ambiente.marche.it/Portals/0/Ambiente/Rifiuti/PW_Traduzione/Pre_waste_waste_prevention_practices_mapping_report%5B1%5D.pdf, accessed 2 May 2018.

Repair Café (2016) "1000 Repair Cafés Worldwide", http://repaircafe.org/en/1000-repair-cafes-worldwide/, accessed 28 October 2016.

Rosner, D. K. (2012) "Devices: On Gender and the Development of Contemporary Public Sites of Repair in Northern California", *Public Culture*, vol 26, no 1, pp. 51–77.

Rosner, D. K. and Ames, M. G. (2014) "Designing for Repair?: Infrastructures and Materialities of Breakdown", *Proceedings of the 17th ACM Conference on Computer Supported Cooperative Work and Social Computing*, Association for Computing Machinery, 15–19 February 2014, Baltimore, MD, USA.

Rosner, D. K. and Turner, F. (2015) "Theaters of Alternative Industry: Hobbyist Repair Collectives and the Legacy of the 1960s American Counterculture", in H. Plattner, C. Meinel and L. Leifer (eds) *Design Thinking Research*, Switzerland: Springer.

RREUSE (2013) *Investigation into the Repairability of Domestic Washing Machines, Dishwashers and Fridges*, www.rreuse.org/investigation-into-the-repairability-of-domestic-washing-machines-dishwashers-and-fridges/, accessed 28 October 2016.

RREUSE (2015) *Improving Product Reparability: Policy Options at EU Level*, www.rreuse.org/wp-content/uploads/Routes-to-Repair-RREUSE-final-report.pdf, accessed 28 October 2016.

Sabbaghi, M., Esmaeilian, B. and Cade, W. (2016) "Business Outcomes of Product Repairability: A Survey-Based Study of Consumer Repair Experiences", *Resources, Conservation and Recycling*, vol 109, pp. 114–122.

Salvia, G., Cooper, T., Braithwaithe, N., Moreno, M. (2015) *What is Broken? Expected Lifetime, Perception of Brokenness and Attitude Towards Maintenance and Repair*, PLATE (Product Lifetimes and the Environment) Conference proceedings, 17–19 June 2015, Nottingham, UK, www.ntu.ac.uk/plate_conference/proceedings/index.html, accessed 28 October 2016.

Scott, K. A. and Weaver, S. T. (2014) "To Repair or Not to Repair: What Is the Motivation?" *Journal of Research for Consumers*, vol 26, no 1, http://jrconsumers.com/Academic_Articles/issue_26/Issue26-AcademicArticle-Scott1-31.pdf, accessed 27 May 2018..

Seyfang, G. and Smith, A. (2007) "Grassroots Innovations for Sustainable Development: Towards a New Research and Policy Agenda", *Environmental Politics*, vol 16, no 4, pp. 584–603.

Shemwell, D. J., Cronin, J. J. and Bullard, W. R. (1994) "Relational Exchange in Services", *International Journal of Service Industry Management*, vol 5, no 3, pp. 57–68.

Slade, G. (2006) *Made to Break: Technology and Obsolescence in America*, Cambridge, MA: Harvard University Press.

Twigg-Flesner, C. (2010) "The Law on Guarantees and Repair Work", in T. Cooper (ed) *Longer Lasting Products: Alternatives to the Throwaway Society*, Farnham: Gower.

Vezzoli, C. and Manzini, E. (2008) *Design for Environmental Sustainability*, London: Springer.

Which? (2011) "Built to Last", www.staticwhich.co.uk/documents/pdf/p20-23_reliability-259732.pdf, accessed 28 October 2016.

Which? (2013) "The Great Extended Warranties Rip-off", http://blogs.which.co.uk/technology/news/great-extended-warranties-rip-off/, accessed 28 October 2016.

White Goods Help (2016) "About Whitegoodshelp", www.whitegoodshelp.co.uk/about-whitegoodshelp-andy-trigg/, accessed 28 October 2016.

WRAP (2012) "Valuing our Clothes", www.wrap.org.uk/content/valuing-our-clothes, accessed 28 October 2016.

WRAP (2016) "The Business Case for Repair Models", www.wrap.org.uk/content/business-case-repair-models, accessed 28 October 2016.

9 ReDress

Maximizing component reuse for fashion

Kim Fraser

Introduction: the long tradition of reuse

Reusing and remaking used clothing is a long-standing practice that began as a way of making the most of valuable resources. Prior to the industrial revolution, creating textiles in Western nations was a labor-intensive process that required specialist knowledge and technique: textiles were valuable possessions both treasured and inherited. This perceived value of clothing textiles was still current in the eighteenth and nineteenth centuries. Ginsburg (1980) construes this apparent value through noting the frequent loss and theft of clothes and textiles that were reported within court cases and newspapers. The importance of clothing is also emphasized by Lemire (2012, p. 147), who specifies textiles as among the most costly of household purchases that benefited from intergenerational use.

Garments were expected to serve a very long life, and they were repaired and remade to enable this intergenerational use – the many layers of garments were constructed simply, with extra long hems and wide seam allowances to allow variations for fit through taking in/up or letting out/down. Along with this, outgrown adults' garments might be cut down and remade for children (Lemire 2012). Often textile components in older garments, worn almost threadbare in a few spots, would be cut up and made into something new or used to repair another. Similarly, patchwork quilts, which were passed down the generations, provided an esteemed avenue for salvaging sections of precious fabrics, once the "life" of the original garment had been exhausted/worn out. Schor (2011, p. 28) discusses the exchange value of apparel at this time as "a primary medium of exchange, second only to metals and precious stones." Lemire (2012, p. 147) further confirms that garments were used as currency indicating that investment in clothing served as a type of savings. She details how garments were given as incentives, passed as wages, gifted, or bequeathed, and sold as second-hand when required. In this manner, second-hand clothes were vitally important in providing "relatively cheap and respectable clothing" to the working classes (Worth 2007, p. 18).

In the early modern era, scarcity of resources was the norm, and most relied mainly on what could be obtained locally. There was universal respect for materials and the specialist labor and techniques involved within their creation. According to Lemire (2004), apparel was a unique exchange commodity, able to store and release value as required. She notes that "[m]uch of the value in garments was founded on the quality of the fabrics used in their construction, the weight, weave, finish and substance of the cloth, plus the presence or absence of braid, lace, buttons or accessories" (Lemire 2004, p. 41). Clothing was "universally owned" and therefore its "value [was] widely understood" (Lemire 2012, p. 148), providing an alternate and accessible currency. This perception thus endorsed the reuse of quality used clothing and emphasized the regard in which textile was held by the majority. Thus, during these times when textile was considered a valuable resource, enduring habits were developed that prolonged its usefulness.

At the dawn of the Industrial Revolution, the efficiencies of creating cloth grew at a rapid rate. The inventions of the carding machine, spinning machine, and power loom, along with improvements in transportation, provided abundant access to resources. The 'shoddy' industry developed to meet the demands of yarn supply. This recycling process recovered fiber from rags which was then spun into yarn and reused by local mills in 'shoddy' cloths. At the same time, ready-made clothing was evolving, and according to Worth (2007, p. 18), shoddy cloth became the "staple of the growing ready-to-wear clothing industry." Thus, although readily affordable, ready-made clothing became synonymous with poor quality and was deemed inferior through the use of lesser-quality fabric combined with poor garment fit. Initially this perception encouraged and endorsed the reusing of quality used clothing, maintaining the widespread regard held for textile.

Lemire (2012) categorizes and contextualizes three historic stages in her discussion of the evolution of the second-hand clothing trade and the changing of the material world (c. 1600–1850). Although the focus of her research was second-hand clothing, essentially her defined stages of "transition from scarcity," "growing abundance," and "industrial plenty" (Lemire 2012 p. 147) aid in contextualizing the changing value of cloth and clothing. With industrialization ("transition from scarcity"), the population shifted to meet the growing work opportunities provided by the textile industry, and according to De Vries (1994, p. 249), a "household-level change" took place. He describes how this "reallocation of the productive resources of households" (De Vries, p. 255) encouraged women and children to enter the paid work force to supplement the household income. In this time of 'growing abundance', Lemire (2012) compares these changes with the change and eventual disdain for the second-hand trade. She discusses how "cash wages became the norm and savings were stored in new savings institutions" (Lemire 2012 p. 153). The implication of 'industrial plenty' was that the respectable middle classes could borrow from banks and no longer had to pawn their clothing (Lemire 2012). Eventually "the purchase of used garments became largely a working-class practice, disdained by most middle-class" (Lemire 2012, p. 153).

The patchwork quilt and shoddy cloth both provide examples of remaking used textiles into new cloth, but more importantly they highlight the significance of the changing attitude held towards the built-in salvage of textile and fiber. But when textile became plentiful and was no longer considered a valuable resource, the long-held enduring practices that prolonged its usefulness were gradually left behind.

Overconsumption and the devaluation of textile

Following 250 years of industrialization the perceived value of textile has drastically changed. The economizing associated with the practice of patchwork relegated it to the resourcefulness of the less affluent and consequently socially undesirable. Similarly, the term 'shoddy' now commonly refers to anything made with inferior material, which directly affects judgment of the shoddy cloth process, reinforcing prejudicial associations. This changing attitude burgeoned in the twentieth century when businesses focused on maximizing profit through increasing production and reducing costs. As standards of living rose in the developed world, the cost of labor became a major focus, and faster, more effective chemical processes were selected to reduce more laborious procedures. Increased efficiency and profit eventuated at the expense of jobs and the environment. In the textile industry, the invention of synthetically made fibers early in the twentieth century created large quantities of fiber without the need for labor-dependent crops. Fiber on demand dramatically changed the textile landscape, an ease of supply devaluing textile for the maker and the user. Later in the century the availability of mass-produced clothing escalated with the advent of low cost, offshore manufacturing.

In 2005, Juliet Schor cited the case of apparel to discuss unsustainable consumption in the global economy. She highlighted how the offshore production of apparel created artificially cheap clothing which in turn contributed to increased purchasing and an excessive accumulation of garments. Apparel buyers exploited the low cost, capitalizing on their economic strength through larger orders far in excess of their needs. These mass stock purchases led to a "pile it high, sell it cheap" clearance strategy, maintaining sales turnover through encouraging the purchase and consumption of multiple items at bargain prices (Farrer and Fraser 2011). Producers in the fast/discount fashion/value clothing sector are now forced to sustain their low costs through depressing wages (Schor 2005) and cost-cutting through utilizing low quality fibers and fast techniques (Fletcher 2014, 2008; Oakdene Hollins 2006). Customers, now motivated by price and status rather than by need, are purchasing far more than ever before (Farrer and Fraser 2011).

When fast fashion developed in the globalized 1990s, widespread overconsumption was already the norm. The fast fashion business model delivers new high-end trends into stores in short regular intervals, at the cheapest possible price-point (Barnes and Lea-Greenwood 2006). This constant newness expands the consumer's wardrobe until, logically, removing the 'old' is necessary: "waste volumes . . . are high and growing in the UK with the advent of fast fashion" (Allwood et al. 2006, p. 2). But the quality of this mass-produced clothing has further deteriorated (in terms of fabric quality and fit) and consequently reuse of these mass-produced garments is becoming a less likely option (Fraser 2013). In this cradle-to-grave approach clothing products are designed from raw materials to meet specific requirements for the market based on appearance, technical performance, and cost. According to SATCol (2007), there generally is little or no attention paid to the consequences of these decisions at end-of-use. They emphasize that while we focus on 'waste', we are missing the point that the original 'design' of the product has defined 'it' as waste, and that we are designing products with a cradle-to-rubbish-bin approach (Fraser 2011).

Textile impact

Textile products are now everywhere, and this ubiquity has consequences. Mass consumption of mass-produced clothing at very low cost provides the consumer with a false understanding of the labor and resources involved (Fraser 2013). Consequently, without a better understanding of further recycling options, 'throwaway' fashion now consigns millions of tons of clothing to landfill. International studies indicate textile waste amounts to between 3 to 5 percent of total municipal waste in the landfill; this equates to 1.8 million tons of textile landfilled in the UK in 2006 (WRAP 2012), 1 million tons in USA (Ecouterre 2012), and 100,000 tons in New Zealand in 2010 (MFE 2011). While these may seem to be a small percentage of total wastes discarded, the quantity per person is far more alarming: 28kg per person per year in the UK, 65lbs or 29kg in USA, and 23kg in New Zealand. To put this further into perspective, 23kg of textile is equivalent to approximately 116 adult sized T-shirts (based on an average T-shirt weight of 200 grams) thrown to the landfill by each and every New Zealand resident in 2010. In terms of total landfill that would be equivalent to 505 million T-shirts disposed in New Zealand annually, within a total population size of less than 5 million (MFE 2011).

In addition to this disturbing wastefulness, textile products create environmental hazards for the landfill that have an adverse effect on humans. Large items of textiles create barriers that behave like pool liners, causing water to pool and stagnate, and at the same time stop water from entering the soil and causing surface water run-off (Garland 2009). Decaying fibers produce toxic gases and leachate (liquid waste in landfill), which contaminates both surface and groundwater sources (Birtwistle and Moore 2007; Waste Online 2017). Additionally, synthetic fibers take many years to decay, releasing toxic gases as they break down, thus prolonging these adverse effects (Draper, Murray, and Weissbrod 2007, p. 5). Also, while natural fibers are considered compostable, surprisingly as woolen garments decompose they produce large amounts of ammonia as well as methane (Birtwistle and Moore 2007; Waste Online 2017).

The processes required to produce virgin fibers is generally considered even more alarming. For example, two of the most prolific fibers in common use are cotton and polyester. Every kilogram of cotton fiber (equivalent to five 200 gram T-shirts) requires 20,000 liters of water (Environmental Justice Foundation 2007) and a kilogram of hazardous pesticide (per hectare) to reach successful harvest. Similarly hazardous for the environment, virgin polyester is extracted from non-renewable crude oil via a chemical reaction (Clark 2004) using very high amounts of energy both in extraction and extrusion of the polymer (American Fiber Manufacturers Association 2016). The process uses large amounts of water for cooling, and the lubricants required can become a source of contamination (Challa 2007). In fact, most synthetic fibers require an extrusion process; however, in addition, nylon manufacture creates nitrous oxide, a major greenhouse gas which is 300 times more damaging than carbon dioxide (EPA 2014). Following the harvesting and extraction of the fiber, further chemical processes and labor are required to process fiber into fabric: from dyeing, weaving, knitting, printing, pre-shrinking, steaming, bonding, and the further finishing treatments. Textile is a water, chemical, and process intensive product.

The textile and apparel supply chain now takes place across huge distances around the globe. The average textile/garment supply chain has a minimum of 50–60 complex procedures and operations. Businesses pass their developing 'product' along the supply chain from one or two specific complex processes, undertaken in a specialized manufacturing facility, to the next specialized facility, for the next specialized process to take place. Gwilt (2011, p. 65) describes how "each phase . . . of design and production will involve different people often working within separate sections of the . . . supply chain, which are at times situated in different geographical locations." In this manner transparency along the supply chain is easily obscured, product passed to the chosen 'contractor' might easily be subcontracted out without the business's knowledge. According to Farrer and Fraser (2011, p. 9), "in reality most businesses 'know' little about their supply and disposal chain . . . and the further they are away from the company headquarters the less they know." It is little wonder then that consumers lack awareness and are confused as to the intrinsic value of textiles, leading to ill-considered disposal.

Thus, if we were to consider that all fiber comes at a cost to people and the environment, then all reclaimed fiber reduces the need for this cost, and therefore has value. There is no doubt that reuse and recycling of textiles would reduce the environmental burden outlined earlier in this chapter (Oakdene Hollins 2006, Fletcher 2008). At first glance textile 'recycling' appears to offer a commercial alternative for dealing with textile waste, because it produces a 'bulk product'. Unfortunately, the process is now considered a down-cycling process that produces an inferior and less expensive cloth. According to Oakdene Hollins (2006), this is due to the heterogeneous mixture of different fiber types and colors extracted from the original discard source. Together with this, recycling negates all value added and embedded along the supply chain. Recycling reduces the clothing item to its elemental value of fiber. In doing so, all laborious intricate detail cut and sewn during the construction of the garment is eradicated. In the same way, all the time and resources consumed during the process of weaving or knitting of the textiles within the garment are destroyed, and any value added through processing the original virgin fibers into yarn is removed. Textile recycling should thus be considered the last step in the lifecycle of a textile product (Finn and Fraser 2013), resorted to after all reuse (both garment and textile) options have been exhausted.

© Fraser 2010

Figure 9.1 ReDress x 5 on rack (2010)
Photo: Kim Fraser.

The *ReDress* case and refashion

The 2009 case study, *ReDress – ReFashion as a solution for clothing (un) sustainability*, explored the possibilities of re-manufacturing fashion clothing as a solution to wasteful fashion consumption and disposal in New Zealand (Fraser 2009). The case is discussed here to draw attention to the potential value to be gained through extending the life of the textile component prior to recycling the fiber, in this way seeking to highlight the opportunities generated by slowing consumerism through a novel form of designing for reuse (Gwilt 2014, p. 141).

The *ReDress* project facilitated the remanufacture of discarded items of clothing that still had structural textile integrity, into 'new' standardized ready-to-wear garments (see Figure 9.1). The purpose of the case was to investigate the barriers for larger scale production of refashioned product, through remanufacturing multiples of a designed *ReDress* prototype from discarded men's trousers. With emphasis placed on the process rather than the artifact, the focus was on repeatability, identifying and recording the appropriate manufacturing knowledge, techniques, and processes required to repeat the making of the product. This, coupled with a comparative analysis of standard manufacturing, identified aspects that were the same or similar in both processes from those that were uniquely different.

The term 'refashion' means to "give new form to something" (Collins English Dictionary 2016). 'Refashion' within the fashion industry has grown into an intervention in the 'take, make, waste' lifecycle of a garment. It is a slow growing, upcycling movement that reuses the textile, maintains the original value added to it, and delivers a new fashion item without the environmental problems required by virgin textile. However, the 'refashion' process has been generally perceived as a 'one-off' domestic process and not practical for commercialization. This 'one-off' perception of the process is a critical issue, since the business model and scale of production for fashion now relies upon a multiplicity of product (bulk product). Through examining the process required for manufacturing multiples of the *ReDress* prototype, it became apparent that perceived barriers to the 'refashion' process primarily concerned the lack of consistency of 'input' stock. The apparent singularity of 'refashion' arises due to the heterogeneous, diversity, and vast nature of new clothing in the fashion market. This diversity directly affects the content of any resulting stocks within the secondary industry: secondary stock is irregular and quantities are unpredictable.

In conventional manufacture of a fashion product, the selection and availability of the cloth is integral to the success of the product. For example, the exact same fabric and trims (down to dye batch number) are sourced in bulk in order to be batch-processed along the specialized supply chain. A vertical retailer in fashion (one that "designs, produces, and sells its own products, without using middlemen or wholesalers" [Collins English Dictionary 2016]) might order thousands of meters of a specific cloth, across a few colorways,

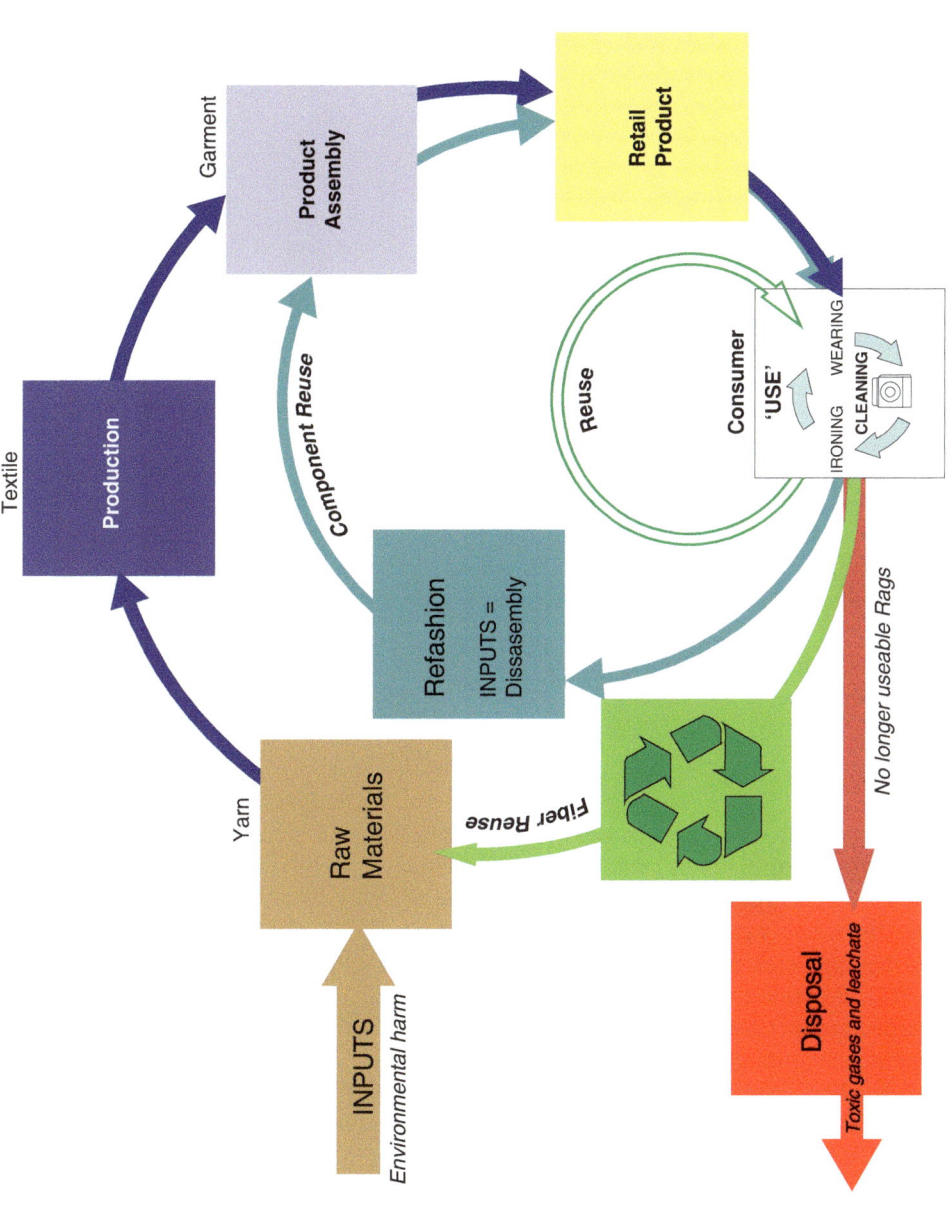

Figure 9.2 Refashion as an intervention in the lifecycle of a garment

Source: Adapted from Fraser (2015).

to manufacture a large quantity in a specific style. The resulting items will be distributed throughout their (global) retail chain, thus dispersing into the consumer population. While the timeframe for selling these units into the population will be known/seasonal, there are no guarantees if, where, or when consumers might return the items to a secondary system. Thus, any hope for commercially refashioning that specific style made from thousands of meters of a specific cloth goes unrealized.

ReDress case findings

In recognition of the input stock dilemma, the *ReDress* project focused on clothing item groups that were noticeable in larger quantity in the New Zealand secondary industry. Men's dress trousers were primarily selected to test the process of refashioning multiples in the *ReDress* case study because they met the criteria of being an appropriate input stock – they had consistency of style, perceived quality of textile, semi-standard shape, similar construction across brands, and limited variation in color. Beneficially, men's dress trousers also featured high quality tailored details that could be reused (fly, pockets, waistband). Additionally, the apparent availability of large quantities of trouser within the secondary system satisfied the 'bulk' requirement to manufacture 'multiples' and potentially commercialize the process. The trouser stock for the *ReDress* project was sourced from several second-hand clothing vendors around New Zealand. At the outset, poor quality textile was assumed to be an impediment to the success of the refashioned product. Therefore, intuitively trouser stocks were selected for their perceived textile quality.

Disassembly

In terms of producing a homogeneous refashioned product for further use, the difficulty of disassembling a garment is considered to be a barrier to effective recycling (SATCol 2007). The *ReDress* case trialed a variety of approaches to the disassembly process. In traditional 'remanufacture', such as that to be found in the appliance or machine industries (Centre for Remanufacture and Reuse 2017), a blanket approach to disassembly is used in which the original discarded product is broken up into its individual components, then (exactly) alike components are grouped as 'stock' items, bundled, and stored. This works in cases where specific components have a limited variety within typical product profiles (e.g. in vehicles, lawn mowers, computers, etc.). However, in the case of clothing, the immense variation of style/ product (fiber, textile, texture, color, pattern, gender, style, size, shape, combined with continual variations in design and construction detailing: collars, cuffs, plackets, linings, facings, fastenings, etc.) decreases any likelihood of gathering random discards together that have exactly the same components. The *ReDress* case revealed that to facilitate

A child's dress cut from an adult's garment which is only partly in good condition.

Figure 9.3 Child's dress recut from adult garment
Source: Resek and Resek (1955, p. 172).

productivity and maximize the input stock, complete disassembly of clothing stock was impractical until specified.

Size limitations

The resulting 'refashion' design is also limited by the restriction of working within the shape of the components formed during first-life construction. Historically, large garments lend themselves more readily to refashioning into smaller garments. For example, in children's clothes (see Figure 9.3), the size of the upcycled child's product is directly related to the two-dimensional area of the available cloth (input stock). That is, larger panel sizes offer more scope to be refashioned. In terms of the *ReDress* project, even disregarding that menswear garments tend to have larger panels than womenswear, the 'usable piece size' in men's trousers is somewhat limiting, as is apparent in Figure 9.4.

While all men's trousers do not each have exactly the same two-dimensional area, they are generally a predictable shape. For example, men's dress trousers are constructed from four main pattern shapes: two front panels and two back panels. The 'back' is usually larger than the front through the seat area and the 'front' will be shorter in the crutch area. At the same time, trouser styles will tend to vary in leg width, which will affect the 'refashion' style. The *ReDress* case revealed that although there is an expected direct relationship between the size of the original trouser and the smaller refashioned output, placement of pattern pieces for 'refashion' is much more complex. Expert consideration must be given to the diversity of the original stock's panel size limitation. Exact placement of pattern pieces must

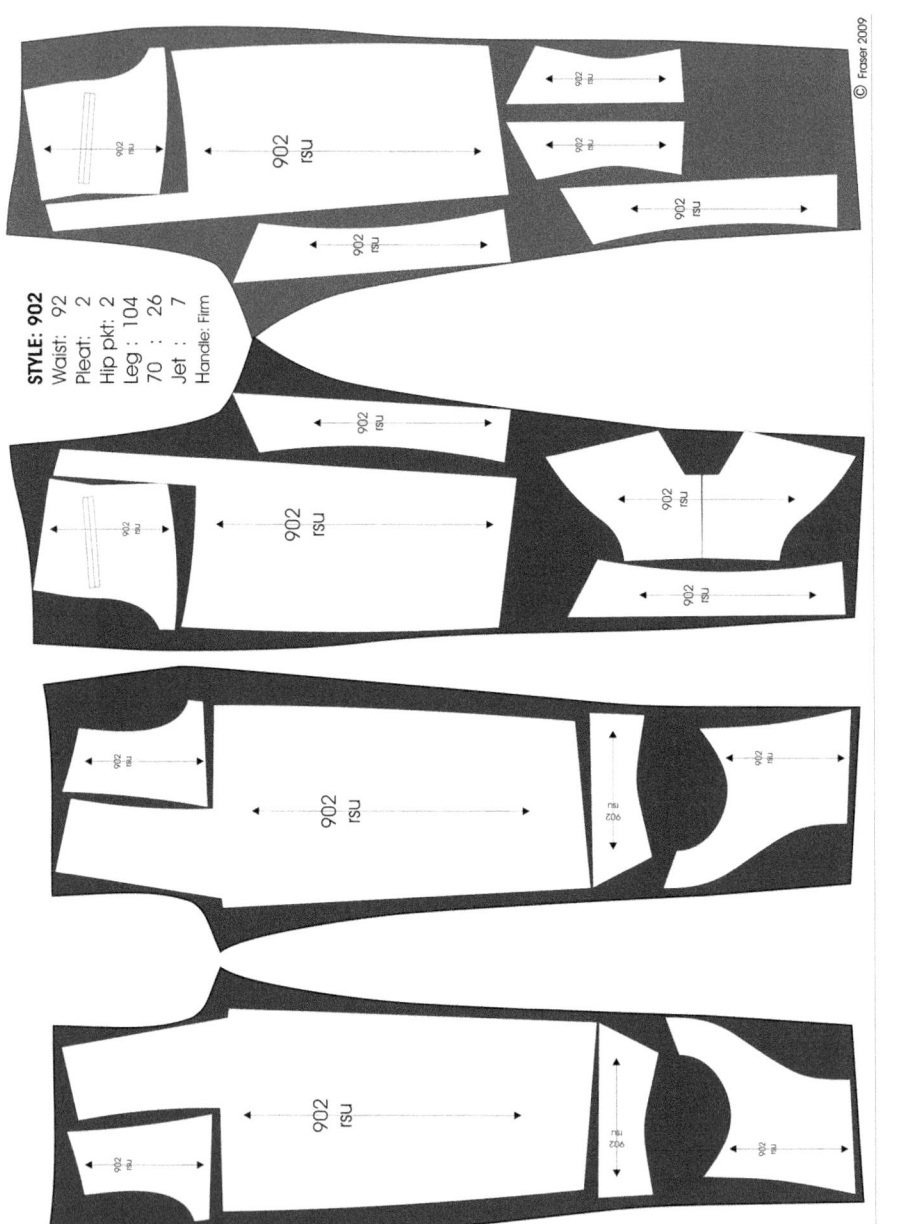

STYLE: 902
Waist: 92
Pleat: 2
Hip pkt: 2
Leg : 104
70 : 26
Jet : 7
Handle: Firm

© Fraser 2009

Figure 9.4 ReDress pattern layout on disassembled trouser (2009)

Photo: Kim Fraser.

consider the original 'grain' and can also be affected by the position of existing details in first life stock.

Component reuse

In the *ReDress* case, each pattern element was designed and cut to purposefully minimize further unnecessary construction through reusing or eliminating existing design details. For example, existing hems and edges that were already cover-stitched were re-utilized as edges, and the reuse of buttonholes, pockets, plackets, waistbands, and zips within the trouser were recreated as openings or details with alternative placement and purpose.

A key discovery in the *ReDress* project was that 'partial disassembly' can provide larger panels for re-cutting from. This reinforced the notion that disassembly requirements may be specific to each refashioned style, and the timing for disassembly is therefore determined by the specifications of the style. The author identified 'component reuse' as the best terminology to describe the 'refashion' process (Fraser 2011). The sizing restriction that necessitates 're-cutting down' to smaller garments is a perceived barrier to 'refashion'. However, through recognizing the disassembly limitation alongside the re-cutting down restriction, a key difference was identified in the approaches, from the cutting process of the original trouser, to the approach in cutting the *ReDress* prototype.

Trouser: The pattern pieces are laid onto the textile within the length and selvedge boundaries. The garment panels are cut away from this allowance. This is a 'reductive approach': garment panels are 'cut away' from the source.

ReDress Prototype: Partially disassembled garment panels are maintained at a maximum surface size and built up where necessary. This is an 'additive approach': garment panels are built up to accommodate the pattern, much like patchwork. For example, in Figure 9.5 the long circular skirts of the 1950s were cut from standard-width fabrics with a small join at the selvedge to supply the extra length.

The key finding in the *ReDress* project was that 'component reuse' for 'refashion' utilizes both an 'additive' and 'reductive' approach. 'Partial disassembly' affords the opportunity to use the 'additive approach'. This approach advantageously allows for refined details, designed and constructed within the original manufacture to be included in this new 'refashion' garment.

Planned variations of style

Pattern-drafting for 'refashion' generally requires a different way of thinking to conventional two-dimensional pattern-drafting. The designer must design from the shapes of the existing garment rather than defaulting to the cutting of predetermined design and pattern shapes. Similarly, cutting 'refashion' pattern shapes from disassembled garment panels also differs from conventional cutting, in that cutting '1 pair' (the same shape mirrored) may not be possible. Panels within the original garment may retain a 'memory' of a previous body

Full Circular Skirt

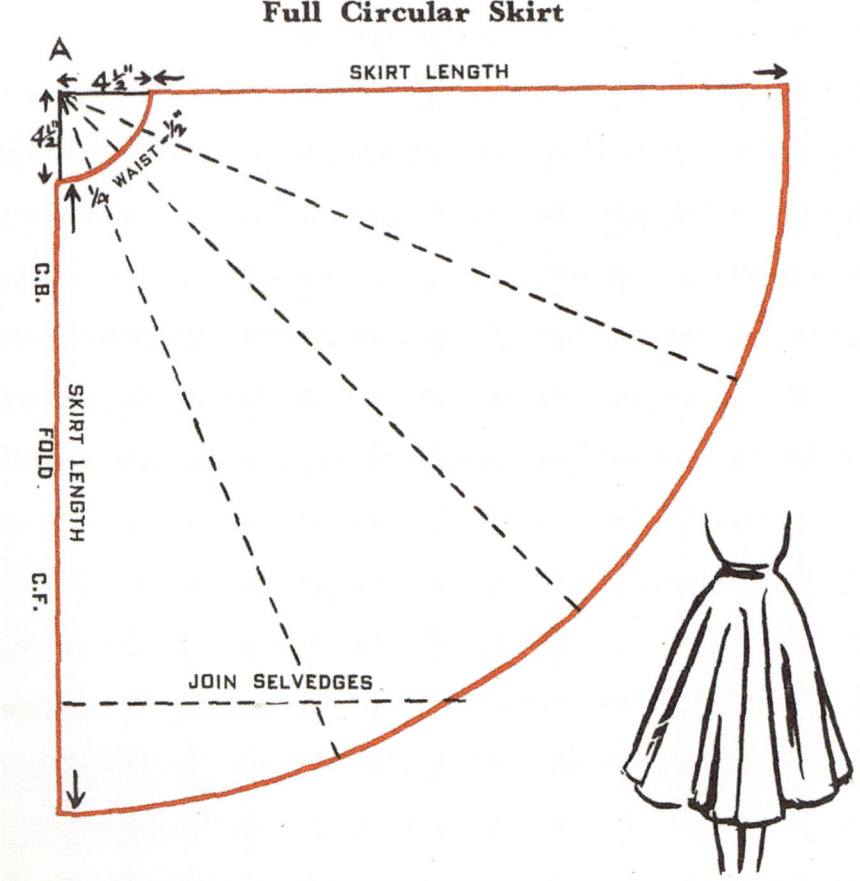

Figure 9.5 1950s full circular skirt pattern

Source: Resek and Resek (1955, p. 36).

and become misshapen: consequently, the distorted panel will no longer mirror the other side. Similarly, positioning of targeted refined details will differ across input stock, rendering some stock items unusable – this is an acknowledged difficulty. However, planning alternative options for these differences during prototyping both minimizes time inefficiency and maximizes stock solutions. For example, during the development of the 'T' series prototype, the only way to fit the new pattern piece on the disassembled panels was to disregard the standard convention of keeping within the straight grain, which maintains the best stability. Choosing instead to adjust the direction of the pattern piece, which produced a better fit on the panel, but caused a diagonal grain.

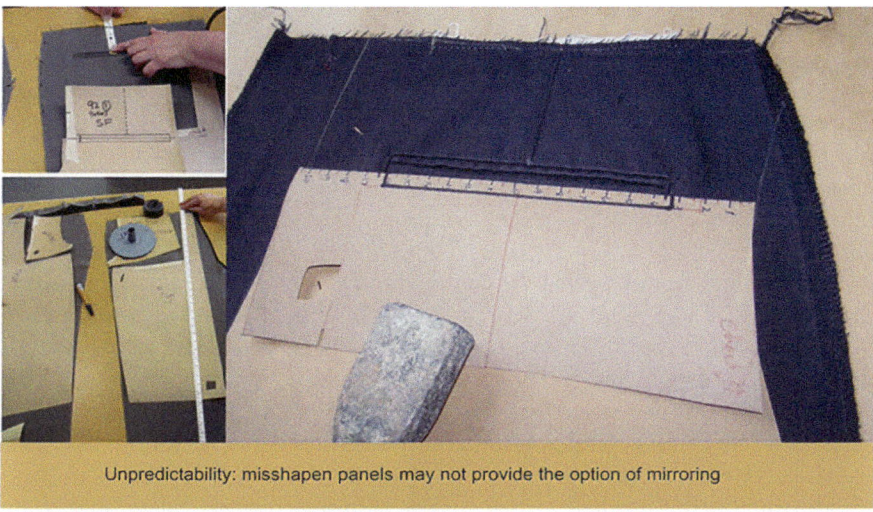

Unpredictability: misshapen panels may not provide the option of mirroring

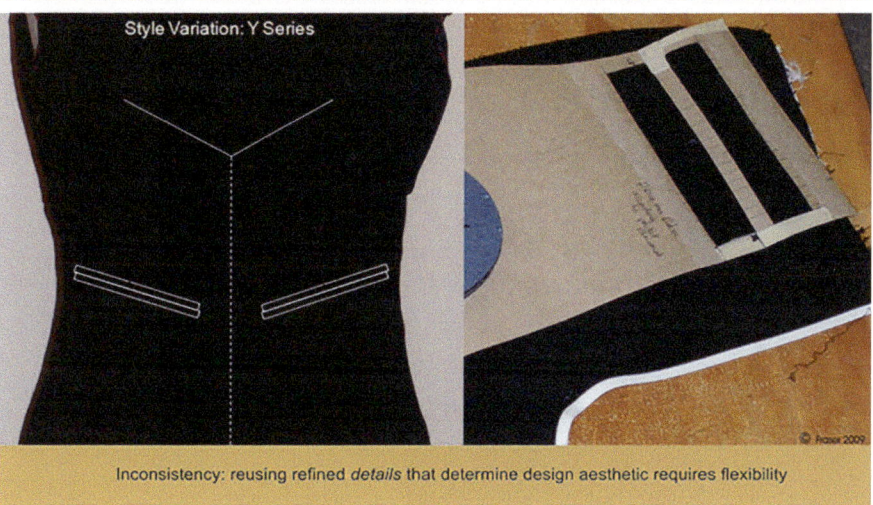

Style Variation: Y Series

Inconsistency: reusing refined *details* that determine design aesthetic requires flexibility

Figure 9.6 Plan for variations of style
Source: Ginsburg (2009). Photo: Kim Fraser.

This resulted in a change to the angle of the pocket placement (see Figure 9.6) and beneficially the 'Y' series was created. This example highlights the need for planned variations of style, which can be adapted to best fit on the diverse dimensions of the disassembled panels, or 'planned variations of style', to allow for inconsistency in first life stock.

Textile durability and integrity

All parts of a garment can be reused, if the garment is reclaimed before it reaches the landfill. While there are a number of style factors that affect the reusability of an unwanted garment, the most prominent factor is the quality of the textile. If the textile has deteriorated, the garment is no longer wearable and the textile is no longer useable. However, it is still possible for the 'fiber' within the textile to be recovered. Table 9.1 identifies the most probable lifecycle path for unwanted garments by their general condition in terms of the quality of textile, style and fit.

It is apparent from this that the foundation for a reuse step already exists in the lifecycle of a garment. In New Zealand, there are plentiful opportunities to donate unwanted clothing through a well-established secondary system – reputable charity shops that accept clothing donations, vast network of clothing banks throughout even the smallest of rural communities, trustworthy online sales through TradeMe®, and good consignment opportunities through established recycle boutiques. However, it should also be mentioned here that there are known issues associated with the global clothing charities. Schor (2002) and Norris (2005) discuss how clothing 'charity' surpluses are shipped to developing countries, to be sold on to shoddy mills or given away by charitable foundations under the guise of 'humanitarian aid'. Norris's (2005) qualitative study confirms that the view that the influx of this seemingly charitable trade, cheap and free clothing, actually undermines local producers, contributing to the erosion of the local garment industry and creating more poverty (see also Poverty Inc 2014).

One of the "Best Practice" findings from WRAP's (2012, p. 10) report on UK clothing confirms that "the largest waste footprint reductions are achieved by extending product lifetime." They suggest designers should "design for durability."

Table 9.1 General conditions for garment reuse (Fraser 2015)

Lifecycle	Condition of textile	Style/fit	Reused as:
Reuse	Good quality textile in good condition	Generally classic style – not trend driven, usually well made, good fit	**Garment** – *as is to be re-worn*
Component Reuse	Satisfactory to good quality textile over majority of garment	May be trend driven style, may require repairs, may be poor fitting style.	**Textile** – *partial deconstruction of garment*
Recycle	Poor quality textile, may be: shabby, pilled, matted, faded, holey, torn, stained, thread-bare	May be trend driven style, may require repairs, may be poor fitting style.	**Fiber** – *elemental or total deconstruction of garment*

Equally Fletcher and Grose (2012) highlighted the need for garments to be made to the highest quality possible to ensure they hold their value and can be resold and reused. 'Component reuse' is a way of extending the life of the textile and its inherent value. But the success of this method is reliant on good quality input stock, and therefore on the quality and durability of the textile. An unexpected finding from the *ReDress* case concerned the high percentage of mass-produced brands in the main trouser stock sourced through the secondary system in New Zealand. But more significant, in terms of 'refashion' and 'reuse', was the poor quality of the textile within this secondary trouser stock. This pervasive presence in New Zealand of low-grade mass produced stock is a significant threat for 'reuse'; shabby stock is less likely to be sold for the purpose of wearing, and poor quality textile is no longer useful for 'component reuse'. Down-cycling, in this case becomes the only intervention available before landfill disposal.

Trend-based fast fashion and mass-produced garments are being developed with a short life expectation. The likelihood is that the style will no longer be reusable at the end of this life. Fast and slow rhythms of clothing use were identified by Fletcher and Tham (2004) where they provide an example of how a 'fast' (trend-based) item might be manufactured from recycled polyester, which can then be recycled again at the end of its life. The *ReDress* case highlights that the size of garment panels is another area for consideration, while designing to provide potential for the next 'component reuse' cycle: small, specifically shaped components are more difficult to recut for another purpose. If the design of the trend-based garment demands a complex fit that necessitates the cutting and joining of specifically shaped smaller pieces to create that fit, then the designer could consider using a lower grade fabric that can be ethically down-cycled at the end of this shortened lifecycle. Therefore, in order to extend these products' lifetime, along with using better quality textiles where possible, designers could also consider eliminating unnecessary seams. Refashioning, once the trend has moved on, will be better enabled, if the component maintains the largest possible area of uncut cloth.

Producer responsibility

In a recent study of post-retail responsibility for garments, Kant Hvass (2014) discussed the responsibility of the producer through industry-driven post-retail initiatives. The carpet industry was one of the first industries in New Zealand to introduce a product stewardship scheme as a method for diverting the large volumes of used carpet from landfill disposal. In 2012 Cavalier Bremworth launched a carpet recycling scheme, using wool carpet to produce a new nonwoven needle punched carpet underlay made from 100 percent recycled wool (Cavalier Bremworth 2016). The company offers the commercial sector an end-of-life take-back opportunity for wool carpet. They encourage residential carpet recycling through an alignment with a waste exchange organization and textile product business in Auckland to recycle the old carpet. The intention is to reduce unwanted residential carpet from being disposed in a landfill and provide a targeted collection point for similar product stock to accumulate as input stock. Unfortunately, just providing access to the web links does not guarantee that the consumer will follow-up with this intervention.

One of the expected problems of using discarded clothing as an input source for manufacture is the heterogeneous nature of the discards. Kant Hvass (2014) discusses Patagonia, Marks & Spencer, Levis, and H&M as examples of how store/brand targeted approaches, or in-store take-back scheme, makes it easier for the individual consumer to return no longer required garments. Similarly, the targeted collection point provides some assurance regarding quantity and quality of garment input stock for the purposes of upcycling. Perhaps a wider uptake of in-store take-back schemes may help to re-establish producer awareness of value and develop a more custodial relationship with their garment products.

Conclusion

The ubiquity of textile has become its downfall. The rock bottom price of textile products obscures their real impacts on the environment. Consumers gain a false understanding of the labor and resources involved, and lack understanding of the intrinsic value of textiles. This devaluation has permeated all textile product, and consequently millions of tons of textiles are now consigned to a landfill globally.

This chapter investigated the long-standing practice of reusing and remaking used clothing within a modern setting, to test the possibility of extending the life of valuable resources and slowing the consumption and wasting of garments. Unfortunately, the inferior quality of mass-produced clothing means it is less desirable as a 'reuse' option. The *ReDress* case study highlighted the benefits of a component reuse process, built into the lifecycle of a fashion product. The case highlighted the potential to produce 'new', standardized, ready-to-wear garments, revealing techniques to manage the perceived haphazard 'refashion' process. 'Refashion', through component reuse, has the potential to extend the life of the textile through creating an additional lifecycle loop with no added environmental cost. As a design intervention, it slows the impacts of accelerated consumerism. By diverting prematurely discarded textile waste from a landfill, 'refashion' can reduce large volumes of textile waste. It can achieve this by replacing the need for the production of virgin textile while maintaining the original value added from the first-life garment, thereby reducing further demand on the environment.

However, 'component reuse' can only take place if the durability and good quality of the textile is considered during the design process. The lifecycle must become a consideration during the design process to enable the 'reuse' of the textile: eliminating unnecessary seams to best enable refashioning of quality fabric; or purposefully selecting lower grade textile for trend-based styles, to permit down-cycling at the end of a shortened lifecycle. This chapter acknowledges that all clothing, with few exceptions, can be reused at some level: garment reuse 'as is' to be worn again, textile reuse through 'refashion', or fiber reuse through recycling. The author argues that both designers and producers need to be aware of designing for reuse, and these principles must be adopted at the time of manufacture. Through considering the new 'refashion' loop the current practices in fashion production will be challenged. Design for reuse could change consumer perspectives and result in the long-term effect of extending the lifecycle of fashion products.

References

Allwood, J., Laursen, S. E., Malvido de Rodriguez, C. and Bocken, N. M. P. (2006) "Well Dressed? The Present and Future Sustainability of Clothing and Textiles in the United Kingdom", www.ifm.eng.cam.ac.uk/uploads/Resources/Other_Reports/UK_textiles.pdf, accessed 25 June 2009.

American Fiber Manufacturers Association (2016) "Manufacturing: Synthetic and Cellulosic Fiber Formation Technology", www.fibersource.com/fiber-world-classroom/manufacturing/, accessed 7 January 2016.

Barnes, L. and Lea-Greenwood, G. (2006) "Fast Fashioning the Supply Chain: Shaping the Research Agenda", *Journal of Fashion Marketing and Management*, vol 103, no 3, pp. 259–271.

Birtwistle, G. and Moore, C. (2007) "Fashion Clothing – Where Does it All End Up?", *International Journal of Retail and Distribution Management*, vol 35, no 3, pp. 210–216.

Cavalier Bremworth (2016) "About", www.cavbrem.co.nz/environment/carpet-recycling. aspx, accessed 17 October 2016.

Centre for Remanufacture and Reuse (2017) "What Is Remanufacturing?", www.remanu facturing.org.uk/what-is-remanufacturing.php, accessed 29 June 2017.

Challa, L. (2007) "Impact of Textiles and Clothing Industry on Environment: Approach Towards Eco-Friendly Textiles", www.fibre2fashion.com/industry-article/1709/impact-of-textiles-and-clothing-industry-on-environment?page=1, accessed 17 January 2016.

Clark, J. (2004) "Polyesters", www.chemguide.co.uk/organicprops/esters/polyesters.html, accessed 7 January 2016.

Collins English Dictionary. (2016) *Collins English Dictionary,* New York: Harper Collins.

De Vries, J. (1994) "The Industrial Revolution and the Industrious Revolution", *The Journal of Economic History*, vol 54, no 2, pp. 249–270.

Draper, S., Murray, V. and Weissbrod, I. (2007) *Fashioning Sustainability: A Review of Sustainability Impacts of the Clothing Industry*, UK: Forumforthefuture, www.forum forthefuture.org/project/fashioning-sustainability/overview, accessed 25 June 2009.

Ecouterre (2012) "Infographic: How Many Pounds of Textiles Do Americans Trash Every Year?", www.ecouterre.com/infographic-how-many-pounds-of-textiles-do-americans-trash-every-year/, accessed 13 January 2016.

Environmental Justice Foundation (2007) *The Deadly Chemicals in Cotton*, Report available at: http://ejfoundation.org/report/deadly-chemicals-cotton, accessed 25 February 2015.

EPA [U.S. Environmental Protection Agency] (2014) "Overview of Greenhouse Gases", http://www3.epa.gov/climatechange/ghgemissions/gases/n2o.html, accessed 7 January 2016.

Farrer, J. and Fraser, K. (2011) "Sustainable "V" Unsustainable: Articulating Division in the Fashion Textiles Industry", *Anti-po-des Design Research Journal*, vol 1, no 4, pp. 1–12.

Finn, A. and Fraser, K. (2013) "Design for Redesign: Can Old Fashioned Strategies Provide New Opportunities for Sustainable Fashion?", Paper presented at Fashion and Social Responsibility Conference, Minnesota.

Fletcher, K. (2008) *Sustainable Fashion and Textile: Design Journeys*, London: Earthscan.

Fletcher, K. (2014) *Sustainable Fashion and Textile: Design Journeys* (2nd ed.), London: Routledge.

Fletcher, K. and Grose, L. (2012) *Fashion and Sustainability: Design for Change*, London: Laurence King Publishers.

Fletcher, K. and Tham, M. (2004) "Lifetimes", www.katefletcher.com/lifetimes/index.html, accessed 18 October 2007.

Fraser, K. (2009) "ReDress: ReFashion as a Solution for Clothing (un) Sustainability", Master's Thesis, AUT University, Auckland. http://aut.researchgateway.ac.nz/handle/10292/817, accessed 2 November 2010.

Fraser, K. (2011) "ReFashioning New Zealand: A Practitioner's Reflection on Fast Fashion Implication", *International Journal of Environmental, Cultural, Economic and Social Sustainability*, vol 73, pp. 275–288.

Fraser, K. (2013) "Throwaway Fashion: The Real Cost of Cheap Fashion for New Zealand", Paper presented at OnSustainability 2013 Conference, Japan, www.youtube.com/watch?v=CBLaqnXFTts&list=PL428534F575A9451A&index=45, accessed 10 May 2015.

Fraser, K. (2015) "ReDress: Reducing Textile Waste Through Component Reuse", Paper presented at UnMaking Waste Conference, South Australia, http://unmakingwaste2015.org/conference-proceedings/, accessed 4 December 2015.

Garland, J. (2009) "Fast Fashion from UK to Uganda", *BBC News*, 20 February 2009, http://news.bbc.co.uk/2/hi/uk_news/7899227.stm, accessed 18 April 2011.

Ginsburg, M. (1980) 'Rags to Riches: The Second-Hand Clothes Trade 1700–1978', *Costume*, vol 14, pp. 121–135.

Gwilt, A. (2011) "Producing Sustainable Fashion", in A. Gwilt and T. Rissanen (eds) *Shaping Sustainable Fashion: Changing the Way We Make and Use Clothes*, London: Earthscan, pp. 59–73.

Gwilt, A. (2014) *A Practical Guide to Sustainable Fashion*, London: Bloomsbury.

Kant Hvass, K. (2014) "Post-retail Responsibility of Garments – A Fashion Industry Perspective", *Journal of Fashion Marketing and Management*, vol 18, no 4, pp. 413–430.

Lemire, B. (2004) "Shifting Currency: The Culture and Economy of the Second Hand Trade in England, c. 1600–1850", in A. Palmer and H. Clark (eds) *Old Clothes, New Looks: Second Hand Fashion*, London: Berg Publishers.

Lemire, B. (2012) "The Secondhand Clothing Trade in Europe and Beyond: Stages of Development and Enterprise in a Changing Material World c. 1600–1850", *Textile*, vol 10, no 2, pp. 144–163.

MFE [Ministry for the Environment] (2011) "Solid Waste Disposal, 2010 Environmental Snapshot", www.mfe.govt.nz/more/environmental-reporting/reporting-act/waste/solid-waste-disposal-indicator/solid-waste-disposal, accessed 11 November 2012.

Norris, L. (2005) "Cast (E)-off Clothing – A Response to K. Tranberg Hansen (AT 20[4])", *Anthropology Today*, vol 21, no 3, p. 24.

Oakdene Hollins Ltd. (2006) "Recycling of Low Grade Clothing Waste", Salvation Army Trading Company Ltd., and Nonwovens Innovation and Research Institute Ltd, www.oakdenehollins.com/pdf/defr01_058_low_grade_clothing-public_v2.pdf, accessed 27 April 2009.

Poverty, Inc (2014) *Secondhand-clothes Undermining Progress*, Vimeo, https://vimeo.com/154615586, accessed 17 October 2016.

Resek, E. and Resek, M. (1955) *Successful Dressmaking*, Melbourne: Colorgravure.

SATCol [Salvation Army Trading Co. Ltd.] (2007) "Memorandum By Salvation Army Trading Co Ltd (SATCol) and the Nonwovens Innovation and Research Institute (NIRI)

Based at University of Leeds", Department of the Environment Food and Rural Affairs, www.parliament.uk/documents/lords-committees/science-technology/st1satradingco.pdf, accessed 21 October 2010.

Schor, J. (2002) "Cleaning the Closet: Toward a New Fashion Ethic", in J. Schor and B. Taylor (eds) *Sustainable Planet – Solutions for the Twenty-First Century*, Boston: Beacon Press, pp. 45–59.

Schor, J. (2005) "Prices and Quantities: Unsustainable Consumption and the Global Economy", *Journal of Ecological Economics*, vol 55, no 2, pp. 309–320.

Schor, J. (2011) *True Wealth: How and Why Millions of Americans Are Creating a Time-Rich, Ecologically Light, Small-Scale, High-Satisfaction Economy*, New York: Penguin.

Waste Online (2017) "Textile Recycling Information Sheet", http://s3.amazonaws.com/zanran_storage/www.wasteonline.org.uk/ContentPages/8789034.pdf, accessed 29 June 2017.

Worth, R. (2007) *Fashion for the People: A History of Clothing at Marks & Spencer*, New York: Berg.

WRAP (2012) "Appendix VI: A Waste Footprint Assessment for UK Clothing", Report, www.wrap.org.uk/sites/files/wrap/Appendix%20VI%20-%20Waste%20footprint%20report.pdf, accessed 19 January 2013.

10 Composting as everyday alchemy

Producing compost from food scraps in twenty-first century urban environments

Vivienne Waller, Linda Blackall, and Peter Newton

Introduction: composting for growing food – the original closed loop system

There is an urgent need to reduce the amount of edible food that we routinely throw away as a staggering one third of the food that is produced is wasted before consumption (Food and Agriculture Organization 2013). Sustainable management of food waste starts with creating as little food waste as possible in the first place, then providing any surplus edible food to community food banks. In this chapter, however, 'food waste' refers to food scraps or food that is no longer edible, in contexts where it is understood to be waste. These are the food scraps that could be reconfigured as a resource through composting, in the original closed loop system.

'Waste' is an ideological category, the necessary flipside of consumerism (Cooper 2008; Cooper 2010). Food waste differs slightly from other waste streams in that some food scraps are an inevitable by-product of eating, regardless of the extent of consumerist manufacture of desire (Campbell 1994). Endless production of any waste is, however, environmentally unsustainable on a planet with finite resources. While consumption "merely uses the world up" (Miller 2012), the model of industrial ecology (Frosch and Gallopoulos 1989) suggests a way to ensure the environmental sustainability of production processes using the concept of closing the loop. Discarded products formerly regarded as waste become the raw materials for another product. For example, excess energy generated from one process can provide energy for another process and thus a closed loop is formed.

This idea of recycling as circular, and hence sustainable, has been critiqued by scholars (Alexander and Reno 2012b; Graeber 2012; Gregson and Crang 2015) who point out that there is always some leakage from the so-called closed loop. With each iteration of recycling, the quality of the material is slightly degraded, and additional energy is usually needed in the process. Moreover, there are often social and environmental hazards left out of the 'good news' story of recycling (Graeber 2012; Gregson and Crang 2015).

While this is the case with regard to the recycling of many manufactured products, such as paper, plastic, and electronic items, composting food scraps and then using that compost to grow food can be a truly closed loop, continuing indefinitely. There is no slow degradation over time, and during the food growing phase, the sun provides energy back into the system through the process of photosynthesis. Of course, the energy seals in the closed loop start to crack when the food scraps are transported long distances to make compost and then further transported to be used in growing food. However, as the examples in this chapter show, recent niche innovations in composting technologies increase the viability of meso-scale on-site composting, requiring no transport of food scraps or compost.

Another complication in the story of closed loop recycling has been highlighted by scholars examining the economies that underpin materials recycling. In order for materials recycling to be viable, it needs to be commercially attractive for manufacturers to recycle the materials into new products (Gregson and Crang 2015), which means there also needs to be demand from consumers for the recycled product (Alexander and Reno 2012a). Here again composting food scraps for growing food can be considered very different from materials recycling. As the examples presented in this chapter show, while industrial-scale composting requires a market for the compost product, distributed systems of onsite composting can bypass the market altogether.

Composting is a natural process that has occurred since life on earth first evolved. During composting, bacteria, fungi, and other microorganisms break down organic materials, such as food scraps, into a nutrient-rich soil-like product called humus. Humans have composted for as long as they have practiced agriculture. For example, 10,000 years ago the Akkadians in Mesopotamia composted, Pliny the Elder refers to the composting practices of the ancient Greeks and Romans, and indigenous Americans are known to have composted (Howard 1943).

In the first half of the twentieth century, several groups around the world 'rediscovered' composting through refining the practice and demonstrating the economic, environmental, and agricultural value of this simple natural activity (Simons 2004). At the same time, however, in industrialized countries, scientists were working on developing artificial chemical fertilizers. There was a need to increase crop yields to meet the growing global food demand, and, as Howard (1943) argued, this coincided with the interests of the explosives manufacturers, who had developed a capacity for the production of nitrogen during the First World War and were looking for new markets. Artificial chemical fertilizers were seen as the way forward, with compost 'scientifically' discredited as a source of nutrients for crops and compost enthusiasts dismissed as cranks (Simons 2004). At the same time, with regard to the management of food scraps, changing attitudes coincided with, and provided a reinforcement for, the emergence and spread of publicly funded waste collection, the provision of the modern landfill (Rathje and Zimring 2012), and the reconfiguration of the citizen as consumer (Smart 2010). Modernist cultural ideals about cleanliness and hygiene created negative attitudes to 'dirt' and 'soil' (Strasser 2000). Food scraps that may have been fed to the chooks or used in home compost were reconfigured as waste for disposal in rubbish bins and eventually the landfill (see Figure 10.1).

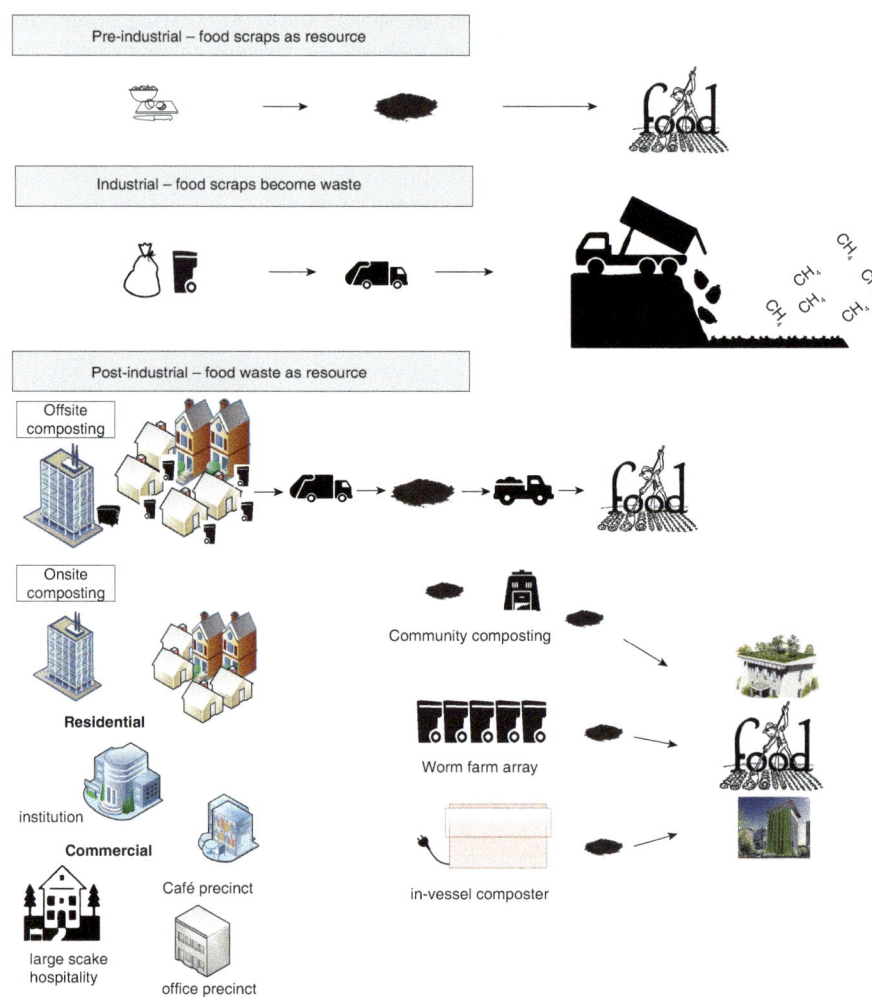

Figure 10.1 Transition pathways for urban food composting

Source: the authors.

The urgent need to reform management of food scraps

The modern landfill has remained the dominant method for managing food scraps in developed countries (Hoornweg and Bhada-Tata 2012). In Australia, for example, approximately eight million tons of food waste were generated in 2008 and, of this, seven million tons went to landfill (Reynolds et al. 2011) at a cost of around $420 million in annual landfill levies.

In developing countries, more than half of the waste generated by households is suitable for composting (Modak et al. 2015). The amount of food waste tends to increase as income increases (Lehmann 2012), and food waste in urban areas worldwide is projected to grow by 35 percent from 2007 to 2025, due to a global increase in incomes combined with population growth (Adhikari et al. 2009).

Sending food scraps as waste to the landfill imposes direct economic and environmental costs with no financial benefit (Adhikari et al. 2009). As well as collection and transportation costs, the economic costs include the cost of the land itself and the costs of building a modern landfill. In 'developing countries', the cost of building a modern landfill is prohibitively expensive. Those landfills that exist tend to be poorly designed and controlled, and open dumping and burning of food waste is still common (Hoornweg and Bhada-Tata 2012). The adverse health effects from open dumping and the burning of waste, including food waste, have been described as a "global health emergency" (Mavropoulos and Newman 2015, p. 6). The main environmental costs of sending food scraps as waste to a landfill relate to the land required, the greenhouse gas emissions produced, and toxic leachate. These problems, which are amplified with open dumping, will be discussed in turn.

It seems that no one wants to live near a landfill, and governments in 'developed' countries are finding it increasingly hard to find land on which to site a landfill to handle city waste. For example, since the late 1990s, New York City has been sending its waste to landfills in neighboring states at a cost of almost one billion US dollars per year (Navarro 2013). The small but densely populated island of Singapore does not have this option, and so has built an island to act as a landfill for 30 years' worth of waste (Catherine Ong Associates et al. 2009). Using land for waste disposal means that it cannot be used to grow food. Adhikari et al. (2009) have calculated that the practice of landfilling, if continually expanded as the dominant means for handling food waste, will have a significant negative impact on world food production capacity.

In landfills, food waste slowly rots to produce methane, a greenhouse gas 34 times more potent than carbon dioxide in relation to its role in global warming and climate change. Greenhouse gas emissions from landfills account for between 3 to 8 percent of a country's greenhouse gas emissions (IPCC 2014; Hoornweg and Bhada-Tata 2012; Adhikari et al. 2009). Rotting food waste also combines with heavy metals in landfills to produce a toxic leachate which becomes a serious problem if it enters the groundwater. Even though engineers may consider that it is possible to construct landfills that do not leak leachate, the problems of methane production from organic waste and space for landfills remain, as well as the amount of open dumping and the number of existing landfills that do produce toxic leachate.

Hence, in both 'developing' and 'developed' countries there is an urgent need to reform food waste management. Recognizing the cost and long term unviability of landfills, some jurisdictions have introduced limits on the amounts of food waste that is allowed to go to landfill (European Commission 2000) or have targets for the diversion of waste from a landfill (for example, Sustainability Victoria 2013). More recently, policy makers realize that by sending waste to the landfill, valuable economic and environmental resources are wasted. In response, some states in the USA have established 'waste to value' policies (Sahajwalla 2013) making it mandatory that all inedible food scraps be composted.

From waste to value: reforming food scraps to use in growing food

Food scraps are a rich source of nitrogen, phosphorus, and potassium, all essential for food production. Phosphorus is a finite resource, but like nitrogen, when leached into the water system from dumps or poorly managed landfills, it is not only lost to the food system, it also becomes a pollutant (Scholz et al. 2013). Compost is an important way of returning the nutrients to the soil that are lost when food is harvested (Lehmann 2012). Hence, applying compost to the soil when growing food reduces or eliminates the need for chemical fertilizers. This also reduces nitrous oxide from fertilizer runoff, nitrous oxide being a greenhouse gas 298 times more potent than carbon dioxide (Forster et al. 2007).

Most agricultural soils in Australia have lost more than 90 percent of their soil carbon due to clearing and cultivation (Norris and Andrews 2010). Adding compost to the soil helps it to store carbon, makes clay soils more friable, and helps sandy soils retain water and nutrients. This means that an application of compost makes plants more resistant to diseases, reducing or eliminating the need for pesticides and herbicides. Less water is needed and higher yields of crops are produced (Department of Primary Industries 2004). In addition, there is less erosion of topsoil and application of compost can decontaminate soil that contains heavy metals.

In the USA, it was found that each ton of food scraps that are diverted from landfill to composting has the potential to reduce approximately one ton of carbon dioxide equivalent (US Environmental Protection Agency 2011). This is without considering the greenhouse gas reductions associated with compost co-benefits such as reduced fertilizer and water use and reduced transport greenhouse gas emissions associated with a shift to on-site composting (as garbage trucks emit carbon dioxide, methane, and nitrous oxide). It has been calculated that if every US city collected and composted food scraps for use in organic farming, the effect on soil carbon storage would offset one fifth of US carbon emissions (Howard 2013). When the composting is combined with growing cover crops, such as rye, the greenhouse gas reductions achieved are estimated to be much higher (Lal 2010; Rodale Institute 2014).

Preliminary evidence also points to community building and health co-benefits from local composting of food scraps (Hyder Consulting 2012). According to a report by the European Commission (2000), composting schemes tend to be popular as they have a "feel-good factor." In a literature review of health impact assessments of composting, Network of Public Health Observatories concluded that "people who engage in composting feel that they are doing their bit for the environment and this in turn increases their sense of wellbeing and health" (Network of Public Health Observatories, 2009, para 19)

There are also economic benefits associated with composting food scraps for growing food. In Australia, high-quality compost retails for around 60 Australian dollars per cubic meter, a figure that represents just part of the economic benefit. Adhikari et al. (2009) have estimated the full economic value of compost, taking into account the value of composted urban food scraps, the fertilizer value, increased yield and resilience of crops, the value of agricultural water saved, reduced landfilling costs, production from land saved from becoming landfill, and reduced greenhouse gas emissions. They estimate that source separation and the on-site treatment of urban food scraps to reduce handling, transportation, and landfilling costs could reduce the total disposal cost of municipal solid waste by almost one third in Europe and the Americas, by approximately two fifths in Asia, and by approximately half in Africa (Adhikari et al. 2009). A recent additional value has been given to compost in countries that have introduced credits for sequestering carbon in soil.

Transitioning to sustainable management of food scraps: the multi-level perspective

While the environmental, health, and economic arguments in favor of composting food scraps for use in growing food are compelling, transitioning from landfill disposal to sustainable management of food scraps involves transforming the entire system of technologies and practices associated with landfilling, a phenomenon that can be characterized as a socio-technical 'regime' (Geels 2011). The socio-technical regime of landfilling involves modern landfill technology, the landfill and waste collection industries, the array of practices and consumer preferences associated with curbside collection and transport of waste, pricing structures, and a supportive regulatory environment, including existing contracts. As Geels (2011) explains, embedded within any socio-technical regime is a given trajectory of development, due to the sunk investments in technologies and systems and the inertia of the existing way of doing things.

In an attempt to theorize the transition to more sustainable ways of living in cities, multi-level perspective scholars have identified two additional analytically distinct levels of practice and context that are relevant to transition processes. These are the broader social, economic, regulatory, and technical 'landscapes', and 'niches', which are innovations that initially operate outside the mainstream (Geels 2011). The theory is that all three levels are involved in transitioning to a more sustainable way of living in cities (Geels and Schot 2007). When changes in the socio-technical landscape exert pressures on the existing regime, this can create what Geels (2002) refers to as a "window of opportunity" to create a change towards a more sustainable (urban) system. This may be through either directly changing the socio-technical regime or introducing niche innovations that coexist with the current regime and lead eventually to a regime change (Geels and Schot 2007; Seyfang and Smith 2007; Kemp et al. 1998).

Most major transition processes are daunting, however, since they involve multiple system transformations and engagement with powerful industry, government, and community groups (regimes) that are commonly resistant to change (Kemp et al. 1998; Geels 2011). The socio-technical landscape of any transition arena comprises multiple stakeholders, attitudes, and practices.

Changes in the socio-technical landscape

In relation to local energy generation and water sensitive precinct design, urban precincts in post-industrial cities are shifting from a total reliance on centralized urban services to distributed systems; for example, incorporating local storm water harvesting and waste water treatment and reuse to augment centralized supply of potable water and distributed renewable energy and storage (Newton 2012). Such transitions occur in response to a set of urban sustainability issues that are becoming increasingly evident in a rapidly growing, urbanizing world. The accelerating pace of consumption (Newton 2011), in the face of planetary boundaries and resource limits, is providing governments with major challenges, including significantly increased rates and volumes of waste generation (Ekstrom 2015).

Diverting food waste from the landfill and establishing robust systems for composting is currently among the most intractable of all urban transitions for several reasons. Foremost among these is the current entrenched practice of disposal via a general purpose 'garbage bin' collection by a centralized service that provides transport to landfill (see Figure 10.1). Alternative processes exist involving food composting as documented in this chapter, but there are a set of 'regime barriers' that must be overcome before transition to some alternative to landfill disposal is possible. On the supply side, there are a range of risks, the foremost being economic and health related. Regarding economic feasibility, industrial-scale composting requires establishing a satisfactory level of demand for the end product as well as an eco-efficient composting process. Although food growers may be initially skeptical, the common experience is that once they have started using good quality compost, supply cannot keep up with demand. Regarding the health risks, many countries have established standards for the operation of composting facilities and for the final product.

The biggest risk is, perhaps, lack of community support. Transitioning to sustainable management of food scraps needs to be socially negotiated. At a minimum, the community needs to play their part in the production of quality compost product through separating their food scraps into a separate bin. Currently in Australia, food scraps constitute roughly half of the domestic waste that goes to landfill (Sustainability Victoria 2013), and most food scraps go to a landfill. Adelaide is a notable exception in that most municipalities are encouraging residents to put food scraps into their green waste bin for removal and subsequent composting for use in farms and vineyards (Waste Management Review 2017).

Studies in Australia and overseas indicate that there are significant contextual and institutional factors that interact with household practices around the management of food scraps. This is seen with distributed energy and water systems, where the nature of the urban fabric, such as urban versus suburban, as well as land use type, represent factors that are proving central to the introduction of different socio-technical innovations, such as solar panels and greywater recycling (Newton and Newman 2013). This would appear to be the case also for the management of food scraps (again, refer to Figure 10.1). For example, worm farm arrays or an in-vessel composter would probably be particularly suited to a high-rise residential environment, a medium density residential complex, or an institution. In low-density suburbs, where detached housing and private yards dominate (Australian Bureau of Statistics 2015), curbside collection and offsite composting may be easier to implement in the absence of on-site (backyard) composting. The demographics of each population are also relevant in terms of preferences and practices.

It is an encouraging sign that residents tend to be more positive about food scraps recycling in municipalities that provide for this process than those in municipalities that do not (Refsgaard and Magnussen 2009). The conclusion is that while establishing appropriate food recycling systems by governments is critical, also required is a shift in resident's subjective identity from that of consumer of waste collection services to active participants enacting the environmentally responsible practice of separating their food scraps (Johnston 2008).

The examples in the next section suggest that this shift may be more achievable when residents are made aware that they are producers of a valuable product that influences the quality of the food they themselves might eat.

Sustainable regimes of managing food scraps

In many post-industrial cities, there is now a window of opportunity for change towards a more environmentally sustainable regime of managing food scraps. The remainder of this chapter adopts a multi-level perspective to present the salient features of three examples of more sustainable management of food scraps. The first example involves a regime change, resulting in the world's largest food scrap collection and composting service. The other examples are of niche developments, with the first of these involving the world's largest municipal waste collection agency.

Source separation, curbside collection, and industrial scale composting

Each day, San Francisco collects 700 tons of food scraps for composting. San Francisco is a city of 800,000 people, almost two thirds of whom live in apartment buildings. It was the first high-density city in the US to undergo a regime change in the management of food scraps, commencing with curbside collection and composting of commercial food scraps in 1996 and expanding to curbside collection and composting of residential food scraps in 1997.[1]

The San Francisco government department responsible for the environment, SF Environment, initiated and oversaw the transition to a sustainable regime for managing food scraps. Its strategic vision is explicitly about sustainability, health, and social equity. Hence their vision regarding waste management is much broader than the focus on diversion from the landfill that is typical of operational agencies charged with waste collection services. This broader vision is reflected in the importance that SF Environment gives to the environmental and health benefits of composting food scraps, and the desire for a quality compost product that can be used to grow food (SF Environment 2016).

In order to facilitate the new regime for managing food scraps, SF Environment deliberately transformed the city's waste collection industry. Rather than periodically put out tenders for the delivery of waste services, the department created a monopoly for the company that collects waste in San Francisco. This has given the company long-term security and the space to continually improve their operations. SF Environment also created financial incentives for this company to meet SF Environment's environmental goals regarding diversion from landfill and compost contamination levels.

In order to be able to produce a quality compost product, it was critical to engage the community and obtain their commitment to actively separating their food scraps, rather than remaining passive consumers of waste collection services. Different sizes and permutations of food scrap bin were trialed, with the results of surveys of consumer satisfaction driving the bin system that was eventually rolled out. Enormous effort was invested in an education and information campaign, including education about how to use newspaper as a kitchen compost bin liner and how to wash the bins. Officers went door to door in apartment buildings, delivering kitchen buckets and working with property managers to set up food separation systems.

SF Environment also made the regulatory environment more supportive of curbside collection and composting. The Mandatory Recycling & Composting Ordinance, passed in 2009, makes it illegal for residents or businesses to place food scraps in the garbage. This is considered to have had a big impact on the separation of food scraps, and by 2014, 80 percent of food scraps were being diverted from the landfill (SF Environment 2009). All businesses are required to provide color-coded, labeled containers in convenient locations with the green container used for food scraps. Another law requires disposable food wares, such as cutlery and plates, to be recyclable or compostable. San Francisco certifies a particular type of compostable bag, which is the only type allowed in the bins. Laws were also introduced requiring new multi-residential developments to have chutes for compostable material conveniently located for all residents as well as areas for compostable materials to be stored and collected. Waste inspectors exist, although their role is more educative than punitive.

In addition, garbage collection practices have changed to accommodate the new regime. Curbside collection has remained, but there is now a weekly collection of food scraps, of recycling, and of landfill garbage. One truck collects the recycling and landfill garbage, and a separate truck collects the food scraps. The food scraps get taken to a composting facility approximately 50 miles away where it is turned into compost of a quality suitable for organic food growing. Market demand drives how the compost is used, but generally half is used for landscaping, erosion control, and organic gardening, while the other half is used on vineyards in California (Recology Organics 2018). Stalls at farmer's markets provide information about the composting process and show samples of compost produced from San Francisco food scraps. This closing of the loop is critical in order for residents to appreciate their role in producing a commercially valuable product from their food scraps, and the importance of properly separating it from the garbage destined for the landfill. Residents are aware that they now play an important part in the city's food production system. As a San Francisco poster promoting the city's composting program claims: "The carrot tops you compost in January could end up becoming part of the carrots you purchase in July."

San Francisco's success has inspired several hundred cities throughout North America, including Toronto, Vancouver, Portland, Minneapolis, and Seattle, to introduce sustainable regimes for managing food scraps through a combination of source separation, curbside collection, and industrial scale composting.

Niche models of managing food scraps – appropriate to high-density built form

The next two examples are each of niche models of composting which are particularly suited to high density living, an urban form increasingly representative of twenty-first century cities. These examples show that it is possible to bypass the need for a market in the production and distribution of the compost product.

Small-scale community composting

Community composting has the advantage that there is no need for expensive curbside collection, no transport costs and associated greenhouse gas production, no odor from bins awaiting collection, and no pollution, noise, or wear and tear on the road from garbage trucks. In addition, it may not be necessary to develop a market for the compost product.

The model of community composting is well developed in New York City (NYC), a city with a population of eight and a half million and the world's largest municipal waste collection agency (NYC Department of Sanitation 2015). NYC is an example of extremely high-density living with over 27,000 people per square mile. In the last 20 years, champions of environmental sustainability have experimented with different ways of managing food scraps, appropriate to high-density living. By 2013, there were more than 200 community composting initiatives in NYC, ranging from community gardens to meso-scale processing of up to two tons per week. These initiatives relied on a combination of government funding from the NYC municipal waste collection agency and volunteer effort.[2] Because community composting is opt-in, those involved have tried to educate about and encourage composting through a suite of programs, compost learning centers and compost demonstration sites, master composter training, and a compost helpline. Although the master composter training and compost helpline are specifically designed to assist with home composting, all of the programs promote the idea of producing compost from food scraps. Indeed, the compost learning centers and demonstration sites are showcases of compost formed from community food scraps (Goldstein 2013).

Home composting is not feasible for most NYC residents due to the high-density apartment living. Community composting encourages residents, who do not have the will or ability to home compost, to be involved in the production of compost through freezing their food scraps and bringing them to one of dozens of staffed drop-off points. These include weekly farmer's markets, community gardens, libraries at designated times, and selected subway stations at morning peak hour (Goldstein 2014). No plastic bags are allowed in the collection. While community gardens tend to use traditional compost bins or heaps, the larger community composting operations use medium size open windrows (long heaps). The resulting compost is used in local urban farming projects and is for sale at farmer's markets, adjacent to the food scrap drop off. It is considered critical to the success of the program that people are actively involved in using the compost to grow food or at least can see the final compost product. In this way, people come to understand first-hand the valuable resource that can be produced from their food scraps and to recognize the significance of their role in the local food production process.

The regulatory environment has been conducive to these niche innovations, as New York regulations allow windrows for medium size static pile composting to be located only 200 meters from residences. Despite the local success of these programs, just a small percentage of the total food scraps in NYC is processed through community composting. By 2013, New York was still sending 1.2 million tons of food scraps as waste to landfills in a year (Navarro 2013). Hence, in June 2013, the mayor of NYC announced the beginnings of a city-wide curbside collection and composting program. Community composters have argued that their continued existence is critical to the success of this regime change, due in particular to their role in promoting the idea of producing compost from food scraps (BioCycle Breaking News 2013). It remains to be seen whether the niche innovations of community composting continue to co-exist with the new regime of food waste management in New York City, although it is certain that their presence was partly responsible for the fact that composting was chosen as the alternative to landfilling food waste.

Community composting, as practiced in New York City, requires a larger commitment from participating households and businesses as they need to actively transport their food scraps to a community garden or food scrap drop off point. By contrast, a different form of community composting was trialed by the City of Yarra in Melbourne, Australia, whereby volunteers on bicycle trailers picked up food scraps from participating cafés and took these to one of several local mid-size compost heaps.[3] Although valuable in promoting the idea of producing compost from food scraps, as the bicycle carts were highly visible, the project was not scalable as it relied on extremely committed volunteers to collect the food scraps and transport these to the compost heap. In addition, it relied on local government funding, and the program ended when the funding ended. As Schot and Geels (2008) argue, however, this type of niche innovation should not be viewed as a failure, as real-world experiments have a crucial place in developing successful niches.

Meso-scale on-site composting

There are a variety of technologies that can be used to produce compost on-site. Most involve compost bins, compost piles, in-vessel composters, or worm farms (see Figure 10.1). As with community composting, onsite composting has the advantage that there is no need for expensive curbside collection or transport, and there is no odor from bins awaiting collection.

This example involves the use of an aerobic, continuous feed, in-vessel composter operating at the University of the Sunshine Coast (USC) in Queensland, Australia (see Figure 10.1). This is a midsize campus with approximately 9,000 on-campus students and 700 staff.[4] USC has estimated that, as a result of composting all of the university's food scraps, they are saving tens of thousands of dollars per year on waste collection costs. At USC, on-campus waste collection is managed by cleaning contractors. USC and the cleaning contractor worked together to move to a more sustainable way of managing food scraps, and this partnership is considered to be key to the success of the project. Getting staff and students at the university to commit to the process of separating their food scraps involved some structural changes. USC removed the under desk bin from each staff office and replaced these with a tiny plastic bin, about the size of a takeaway coffee cup. All university staff is responsible for emptying their own tiny bin into the larger bins placed in communal areas. This has reduced the load of the cleaning staff as they no longer have to do the bending, lifting, and changing of bin liners that was part of emptying the under-desk bins. The catering at USC is all run by the university itself, so it was a relatively simple matter to ensure that all of the take-away coffee cups and lids, cutlery, and plates were compatible with the in-vessel composter.

Cleaning staff empties the compost stations every day, washing and disinfecting the lids. A designated cleaner spends three hours per day working exclusively with the in-vessel composter. This individual manually separates used cutlery, plates, coffee cups, and lids and uses a garden chipper to shred these, and also manually removes contamination from the food scraps. This sorting is critical because, despite the visual material and labels on the food bin, both staff and students make mistakes, for example, wrapping apple cores in cling film before putting them in the compost bin. It should be noted that the local council sends the municipality's food scraps as waste to landfill, so there is no local discourse about producing compost from food scraps.

The shred and food scraps are fed into the composter each day via an electric bin lifter. The composter uses a small amount of electricity, producing compost in about two weeks. The resulting compost, which ejects out the other end, is mixed with some soil and used on the university vegetable garden.

As is evident from this description, the management of any in-vessel composter that is larger than domestic size needs to be built into the job description of a specially-trained staff member or team. This, in conjunction with the fact that there is no local discourse about producing compost from food scraps presents a particular challenge. It seems that not all students and staff are aware of their role in producing a valuable product. Without a shift in their knowledge and understanding, contamination of food scraps is inevitable.

However, not all on-site composting needs to be managed by specially trained people. For example, arrays of modular worm farms make use of a technology that can be managed by anyone. In this case, anyone on-site can be involved directly in the composting process and, in return, obtain the benefit of worm castings and worm juice. Whatever technology is used for on-site composting, in urban environments the resulting compost can bypass the usual market mechanisms either by being used on-site or going to local community gardens, local rooftop farms, or community-supported agriculture (Smaje 2014).

Challenges for managing food scraps through composting

This chapter has shown that in more advanced waste regimes, such as those operating in Adelaide, San Francisco, and New York City, consumers of waste collection services are being encouraged to become involved in the food production system through separating their food scraps for the production of compost for growing food locally. Throughout the world, cities are establishing programs to enable the composting of all food scraps. This is, however, not without its challenges.

In locations where composting has been poorly implemented in the past, negative perceptions remain. Badly managed home compost heaps can smell, attract rodents, and actually contribute to greenhouse gas emissions. Similarly, the introduction of large-scale composting in the past has often failed. It has been expensive, produced a low-quality contaminated product, and generated odors unacceptable to the community. Many large-scale composting operations have failed because they have separated the compostable waste from the non-compostable waste after curbside collection, attempting to bypass the need for community involvement in the production process (Adhikari et al. 2009). This post-collection separation process is expensive and invariably results in a low-quality product with high contamination levels. Now it is generally accepted that in order for composting to be environmentally effective and cost-effective, the food scraps need to be separated by the community at source (European Commission 2000; Adhikari et al. 2009). With regard to odor, there are numerous examples of large-scale composting ventures that were unable to successfully manage associated odors and were shut down as a result (Epstein 2011). As the examples in this chapter demonstrate, these issues can be resolved with a combination of citizen involvement, new technologies, better systems and better management. However, the negative perceptions about composting remain, particularly in locations where composting has been unsuccessfully implemented in the past.

There are also challenges to composting from less environmentally friendly and less cost-effective options for managing food scraps. These include burning food scraps for energy, dehydration, and converting food scraps to plastic (for example, see Bernstad and la Cour Jansen 2012; Girotto et al. 2015). Although these options do not realize the full potential value of food scraps, there are vocal industries promoting these as solutions. They can be attractive to decision-makers who have experienced examples of poor implementation of composting, or are not aware of the successful industrial and meso-scale implementation of composting of food scraps in urban environments.

These policy challenges to composting can be overcome through demonstration of successful examples of industrial and meso-scale composting in urban environments. Also required is a rupture of the link between consumer culture and food waste. As this chapter has demonstrated, this link can be broken by reconfiguring food waste as a valuable resource. This also involves a shift in people's subjective identity from consumers of waste collections to active participants in the production of compost.

Conclusions

As outlined in this chapter, the achievements of large, high density cities prove the success of managing the problem of food waste by involving residents in reforming their food scraps into the valuable product of compost. In addition to offsite composting, onsite composting would appear to have considerable potential for being scaled up as a new distributed form of food scraps management in higher-density living situations and precincts. Common to the success of the cases described here is a supportive regulatory environment and source separation of food scraps. Successful separation of food scraps at source relies on a shift in people's subjective attitudes and actions in relation to food scraps. Rather than being passive consumers of waste collection services they need to become active participants in the production of quality compost from their food scraps, recognizing that this compost can be used to grow food locally.

This can have implications for household food provisioning practices and food production practices. Supermarkets, driven by the need to maximize sales, encourage consumers to buy more than they need (Johnston 2008), contributing to the huge amount of edible food that consumers routinely throw away, referred to at the beginning of this chapter. It is conceivable that a tightening of the loop between food production and food consumption may encourage people to consume more local unprocessed foods, acquired through places other than the supermarket, such as local farmers markets and community supported agriculture, and that this in turn reduces the amount of food scraps produced in the first place (Smaje 2014). It is also likely that less pesticides, herbicides, and fertilizers are needed to be used in food production.

More research is needed on these aspects and on how the different approaches to onsite composting compare with off-site composting in terms of greenhouse gas reductions, compost quality and people's engagement with alternative systems for collecting and processing food scraps. It is not to be expected that one size fits all. Particular approaches to reforming food waste management are likely to be more appropriate to particular communities or particular urban forms and fabrics.

Further applied research can road-test different approaches to composting and subsequent food growing, in different types of urban environments, in order to help reshape the future of urban food waste management, as well as urban food production, into a more environmentally sustainable one. The examples in this chapter suggest that, in order for these approaches to also be socially sustainable, it is necessary to subvert the category of food waste, reconfiguring food scraps as a resource. In doing this, it is also necessary to successfully shift peoples' subjective identity from being passive consumers of food waste collection services to active participants in the food production process.

Acknowledgments

The authors would like to thank Jack Macy of SF Environment, Paul Camilleri and Joy Dillon of University of the Sunshine Coast, and the many community composters in New York who all generously gave of their valuable time to discuss composting. Any errors are the responsibility of the authors.

Notes

1 Unless stated otherwise, the data on San Francisco is sourced from observations and interviews with staff of SF Environment. These were conducted in June 2013 by the lead author.
2 Unless specified otherwise, the data on New York is sourced from observations of community composting in New York and interviews with New York community composters. These were conducted in June 2013 by the lead author.
3 The data on City of Yarra is drawn from the lead author's observations.
4 The data on USC is sourced from a presentation by Paul Camilleri and Joy Dillon at the 14th International Australasian Campuses Towards Sustainability Conference (ACTS), Hobart, 5–7 November 2014, and subsequent conversations with the presenters.

References

Adhikari, B., Barrington, S. and Martinez, J. (2009) "Urban Food Waste Generation: Challenges and Opportunities", *International Journal Environment and Waste Management*, vol 3, no 1/2, pp. 4–21.
Alexander, C. and Reno, J. (2012a) *Economies of Recycling: The Global Transformation of Materials, Values and Social Relations*, London: Zed Books.
Alexander, C. and Reno, J. (2012b) "Introduction", in C. Alexander and J. Reno (eds) *Economies of Recycling: The Global Transformation of Materials, Values and Social Relations*, London: Zed Books, pp. 1–34.
Australian Bureau of Statistics (2015) *3218.0 – Regional Population Growth, Australia, 2013–14*, Canberra: Australian Bureau of Statistics.
Bernstad, A. and la Cour Jansen, J. (2012) "Review of Comparative LCAs of Food Waste Management Systems – Current Status and Potential Improvements", *Waste Management*, vol 32, no 12, pp. 2439–2455.
BioCycle Breaking News (2013) "Why New York City Community Composters Are Critical to City-Wide Implementation", www.biocycle.net/2013/06/27/why-new-york-

city-community-composters-are-critical-to-city-wide-implementation/, accessed 27 March 2016.

Campbell, C. (1994) "The Desire for the New: Its Nature and Social Location as Presented in Theories of Fashion and Modern Consumerism", in R. Silverstone and E. Hirsch (eds) *Consuming Technologies: Media and Information in Domestic Spaces*, London: Routledge, pp. 48–66.

Catherine Ong Associates, Chong, C. and Song, F. (2009) "Semakau Landfill", www.waste-management-world.com/articles/2009/03/semakau-landfill.html, accessed 11 February 2015.

Cooper, T. (2008) "Challenging the 'Refuse Revolution': War, Waste and the Rediscovery of Recycling 1900–50", *Historical Research*, vol 81, no 214, pp. 710–731.

Cooper, T. (2010) "Recycling Modernity: Waste and Environmental History", *History Compass*, vol 8/9, pp. 1114–1125.

Department of Primary Industries (2004) "Compost for Vegetable Growers, Fact Sheet 2: Why Use Compost?", www.sustainability.vic.gov.au, accessed 27 March 2016.

Ekstrom, K. (ed) (2015) *Waste Management and Sustainable Consumption*, London: Routledge.

Epstein, E. (2011) *Industrial Composting: Environmental Engineering and Facilities Management*, Boca Raton: CRC Press, Taylor & Francis.

European Commission (2000) *Success Stories on Composting and Separate Collection*, Luxembourg: Office for Official Publications of the European Communities; European Commission.

Food and Agriculture Organization (2013) *Food Waste Footprint: Impacts on Natural Resources*, Summary Report FAO.

Forster, P., et al. (2007) "Changes in Atmospheric Constituents and in Radiative Forcing", in S. Solomon et al. (eds) *Climate Change 2007: The Physical Science Basis. Contribution of Working Group I to the Fourth Assessment Report of the Intergovernmental Panel on Climate Change*, Cambridge; New York: Cambridge University Press.

Frosch, R. A. and Gallopoulos, N. E. (1989) "Strategies for Manufacturing", *Scientific American*, vol 261, no 3, pp. 144–153.

Geels, F. W. (2002) "Technological Transitions as Evolutionary Reconfiguration Processes: A Multi-Level Perspective and a Case-Study", *Research Policy*, vol 31, no 8/9, pp. 1257–1274.

Geels, F. W. (2011) "The Multi-level Perspective on Sustainability Transitions: Responses to Seven Criticisms", *Environmental Innovation and Societal Transitions*, vol 1, no 1, pp. 24–40.

Geels, F. W. and Schot, J. (2007) "Typology of Sociotechnical Transition Pathways", *Research Policy*, vol 36, no 3, pp. 399–417.

Girotto, F., Alibardi, L. and Cossu, R. (2015) "Food Waste Generation and Industrial Uses: A Review", *Waste Management*, vol 45, pp. 32–41.

Goldstein, N. (2013) "Community Composting in New York City", *Biocycle,* vol 54, no 11, p. 22.

Goldstein, N. (2014) "Greenmarkets Facilitate Food Scraps Diversion in NYC", *Biocycle*, vol 55, no 2, p. 20.

Graeber, D. (2012) "Afterword: The Apocalypse of Objects – Degradation, Redemption and Transcendence in the World of Consumer Goods", in C. Alexander and J. Reno (eds) *Economies of Recycling: The Global Transformation of Materials, Values and Social Relations*, London: Zed Books.

Gregson, N. and Crang, M. (2015) "Waste, Resource Recovery and Labour: Recycling Economies in the EU", in J. Michie and C. Cooper (eds) *Why the Social Sciences Matter*, London: Palgrave Macmillan.

Hoornweg, D. and Bhada-Tata, P. (2012) *What a Waste: A Global Review of Solid Waste Management*, Washington, DC: World Bank.

Howard, A. (1943) *An Agricultural Testament*, Oxford: Oxford University Press.

Howard, B. (2013) "How Cities Compost Mountains of Food Waste", http://news.nation algeographic.com/news/2013/06/130618-food-waste-composting-nyc-san-francisco/, accessed 26 September 2015.

Hyder Consulting (2012) *Food and Garden Organics: Best Practice Collection Manual*, Canberra: Department of Sustainability, Environment, Water, Population and Communities.

IPCC (2014) *Climate Change 2014: Impacts, Adaptation, and Vulnerability. Part B: Regional Aspects*, Contribution of Working Group II to the Fifth Assessment Report of the Intergovernmental Panel on Climate Change, Cambridge; New York: Cambridge University Press.

Johnston, J. (2008) "The Citizen-Consumer Hybrid: Ideological Tensions and the Case of Whole Foods Market", *Theory and Society*, vol 37, pp. 229–270.

Kemp, R., Schot, J. and Hoogma, R. (1998) "Regime Shifts to Sustainability Through Processes of Niche Formation: The Approach of Strategic Niche Management", *Technology Analysis & Strategic Management*, vol 10, no 2, pp. 175–198.

Lal, R. (2010) "Managing Soils and Ecosystems for Mitigating Anthropogenic Carbon Emissions and Advancing Global Food Security", *BioScience*, vol 60, no 9, pp. 708–721.

Lehmann, S. (2012) "The Metabolism of the City: Optimizing Urban Material Flow Through Principles of Zero Waste and Sustainable Consumption", in S. Lehmann and R. Crocker (eds) *Designing for Zero Waste: Consumption, Technologies and the Built Environment*, Oxford: Earthscan.

Mavropoulos, A. and Newman, D. (2015) *Wasted Health: The Tragic Case of Dumpsites*, Vienna: International Solid Waste Association (ISWA).

Miller, D. (2012) *Consumption and its Consequences*, Cambridge: Polity Press.

Modak, P., Wilson, D. and Velis, C. (2015) "Waste Management: Global Status", in D. Wilson (ed) *Global Waste Management Outlook*, Antwerp: United Nations Environment Programme and International Solid Waste Association.

Navarro, M. (2013) "Bloomberg Plan Aims to Require Food Composting", *The New York Times*, 13 June, p. A1.

Network of Public Health Observatories, 2009. *Health impacts of recycling biodegradable materials*, West Midlands: West Midlands Public Health Observatory, available at http://webarchive.nationalarchives.gov.uk/20170106195556/http://www.apho.org.uk/resource/view.aspx?RID=78846 Accessed 24 May 2018.

Newton, P. (ed) (2011) *Urban Consumption*, Melbourne: CSIRO Publishing.

Newton, P. (2012) "Liveable and Sustainable? Socio-Technical Challenges for 21st Century Cities", *Journal of Urban Technology*, vol 19, no 1, pp. 81–102.

Newton, P. and Newman, P. (2013) "The Geography of Solar Photovoltaics (PV) and a New Low Carbon Urban Transition Theory", *Sustainability*, pp. 2537–2556.

Norris, D. and Andrews, P. (2010) "Re-coupling the Carbon and Water Cycles By Natural Sequence Farming", *International Journal of Water*, vol 5, no 4, pp. 386–395.

NYC Department of Sanitation (2015) "About DSNY", http://www1.nyc.gov/assets/dsny/site/about accessed 24 May 2018

Rathje, W. L. and Zimring, C. A. (2012) *Encyclopedia of Consumption and Waste the Social Science of Garbage*, Thousand Oaks, CA: Sage.

Recology Organics (2018) "Jepson Prairie Organics" https://www.recology.com/recology-vacaville-solano/jepson-prairie-organics/ accessed 24 May 2018.

Refsgaard, K. and Magnussen, K. (2009) "Household Behaviour and Attitudes with Regard to Recycling Food Waste – Experiences from Focus Groups", *Journal of Environmental Management*, vol 90, no2, pp. 760–771.

Reynolds, C., et al. (2011) "An Introduction to the Waste Input Output Model: A Methodology to Evaluate Sustainable Behaviour Around (Food) Waste", in P. E. Roetman and C. B. Daniels (eds) *Creating Sustainable Communities in a Changing World*, Adelaide: Crawford House Publishing.

Rodale Institute (2014) "Regenerative Organic Agriculture and Climate Change: A Down – to – Earth Solution to Global Warming", https://rodaleinstitute.org/assets/RegenOrgAg ricultureAndClimateChange_20140418.pdf.

Sahajwalla, V. (2013) "Understanding the 'Waste to Value' Proposition," https://www.insidewaste.com.au/general/news/1003439/understanding-waste-value-proposition accessed 24 May 2018.

Scholz, R. W., Uhlrich, A. E. and Eilittä, M. (2013) "Sustainable Use of Phosphorus: A Finite Resource", *Science of The Total Environment*, vol 461–462, pp. 799–803.

Schot, J. and Geels, F. W. (2008) "Strategic Niche Management and Sustainable Innovation Journeys: Theory, Findings, Research Agenda, and Policy", *Technology Analysis & Strategic Management*, vol 20, no 5, pp. 537–554.

Seyfang, G. and Smith, A. (2007) "Grassroots Innovations for Sustainable Development: Towards a New Research and Policy Agenda", *Environmental Politics*, vol 16, no 4, pp. 584–603.

SF Environment (2009) San Francisco Department of the Environment. Mandatory Recycling and Composting Ordinance, http://sfenvironment.org/sites/default/files/policy/ sfe_zw_sf_mandatory_recycling_composting_ord_100–09.pdf, accessed 24 October July 2016.

SF Environment (2016) San Francisco Department of the Environment, About SF Environment. San Francisco: San Francisco Department of the Environment, http://sfenviron ment.org/about, accessed 24 October July 2016.

Simons, M. (2004) *Resurrection in a Bucket: The Rich and Fertile Story of Compost*, NSW: Allen & Unwin.

Smaje, C. (2014) "Kings and Commoners: Agroecology Meets Consumer Culture", *Journal of Consumer Culture*, vol 14, no 3, pp. 365–383.

Smart, B. (2010) *Consumer Society: Critical Issues and Environmental Consequences*, London: Sage.

Strasser, S. (2000) *Waste and Want: A Social History of Trash*, New York: Henry Holt.

Sustainability Victoria (2013) Draft Statewide Waste and Resource Recovery Infrastructure Plan for 2013–2043.

US Environmental Protection Agency (2011) Reducing GHGs through Recycling and Composting US Environmental Protection Agency.

Waste Management Review (2017) "Grants to boost household food scrap recycling" at http://wastemanagementreview.com.au/grants-boost-household-food-scrap-recycling/ accessed 24 May 2018

11 Reuse in earthship construction

Reclaiming the past to shape the future

Keri Chiveralls

Our fragile planet cannot tolerate a growing population of affluent consumers. Technology and 'free markets' will not save us. We cannot wait for governments to lead the way. We must build the new world ourselves, within the shell of the old. Together we can spark a cultural conversation about these issues and creatively explore them in our own context.

(Osmond and Alexander 2016, 1:13:47)

Introduction

This chapter explores reuse in Earthship construction as a form of counter-hegemonic architecture that challenges the "culture-ideology of consumerism" (Sklair 2010) by literally and metaphorically building new physical and social structures from the 'ruins' of industrial modernity (Dawdy 2010). Earthships are a kind of architecture or 'Biotecture,' which is designed and promoted as a 'radically sustainable' approach to building (Earthship Biotecture 2015a), involving the construction of 'off-grid' housing utilizing earth and other materials, including used cans, bottles, and tires, which are conventionally viewed as 'waste'. While the benefits and appeal of Earthships far exceed this one design component, the building with waste aspect has generated much interest and debate. The life-cycle benefits of building with such materials have been points of contention both within and outside the Earthship movement. For example, some have argued it may even be detrimental in the long-term to be building with resources such as aluminum cans, which could be returned to the recycling stream to reduce the need for mining of materials (Hewitt and Telfer 2012, p. 49). However, informed by participant observation and fieldwork conducted at the Earthship Ironbank workshops in South Australia, I propose that the significance of this element of Earthship construction extends far beyond conventional understandings of material lifecycle impacts. Rather, building with waste, results not only in the building of new physical structures from discarded materials of industrial society, but also challenges and confronts dominant cultural and social paradigms and paves the way for the building of new cultural and social systems "within the shell of the old" (Osmond and Alexander 2016 1:13:47). Furthermore, building with waste, particularly bottles and cans, enables the social structures and relationships formed throughout the building process to become symbolically embedded in the building, achieving a sense of permanence from impermanence, and positioning the transient experience of the workshops within the context of larger social and physical structures. This chapter considers reuse in Earthship building as a cultural practice, affiliated with permaculture and 'natural', 'eco', or 'alternative' building movements, which challenges dominant paradigms of mass-production and hyper-consumption (Lipovetsky 2011), and presents renewed possibilities for re-imagining ways of living with the earth and with one another (Harkness 2009).

Background: earthships as counter-hegemonic architecture?

This chapter is informed by participant observation that took place during a series of workshops at Earthship Ironbank (see Figure 11.1 and Figure 11.2) in the Adelaide Hills, over a period of several years. During the Easter and summer of 2014, approximately 60 people from around Australia and the world traveled to Ironbank to participate in workshops. Most participants were involved for a period of one to five weeks, many camping on site. While many applied for, and paid to attend the workshops, there were also other volunteers and visitors who were involved on a more sporadic basis (with some attending for a day or two at a time), along with a more experienced 'crew' who were paid to assist with

Figure 11.1 Earthship Ironbank from below

Photo: Keri Chiveralls.

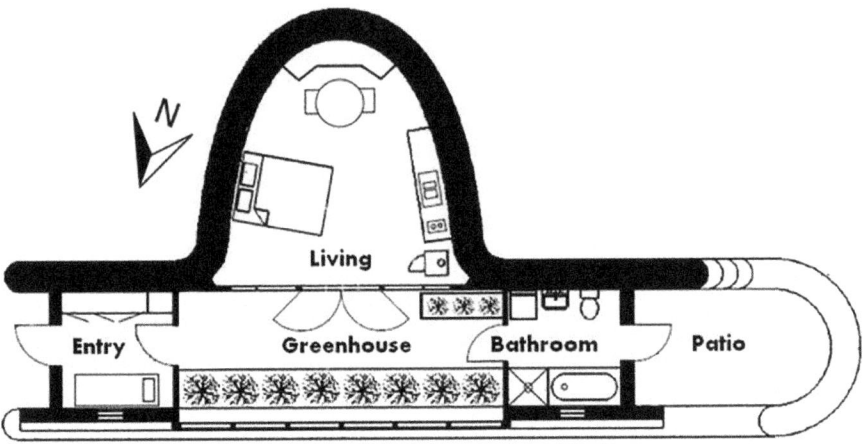

Figure 11.2 Earthship Ironbank floor plan
Image: Martin Freney.

the running of the workshops. Observing and participating in these workshops enabled me to experience reuse in Earthship building firsthand, though at time of writing, I remain very much in the early stages of the journey from neophyte to experienced practitioner, or from 'apprentice' to 'wizard' as the 'Earthshippers' (Harkness 2009) from Ironbank would more likely describe it. The term 'wizard' was employed at the workshops to indicate a high level of skill in Earthship or other mastery, no doubt inspired by Earthship Biotecture creator Michael (Mike) Reynolds' (1989) publication of *A Coming of Wizards*, which tells the story of how he received the inspiration and knowledge to develop the Earthship concept.

Sklair (2010; Sklair and Gherardi 2012) describes "iconic architecture" as a hegemonic project of the transnational capitalist class that "reinforces the culture-ideology of consumerism." Conversely, I consider Earthship construction as a form of counter-hegemonic architecture, in that it challenges prevailing ortho-doxies, categories, and ideas, including "the culture-ideology of consumerism" (Sklair 2010). In her ethnographic doctoral study of the Earthship movement in Taos, New Mexico and Fife, Scotland, Harkness (2009) argues that Earthship building can be seen as a hybrid form of creation that both imagines and practi-cally demonstrates (while continually co-creating) a new (old) way of 'dwelling' in the world that disconnects from and subverts the consumerist paradigm, while re-ordering relationships between people and their environment, one another, and the objects they create. My research conducted at Earthship Ironbank in South Australia supports these arguments, demonstrating how reuse in Earthship build-ing can be viewed as both a "destructive" and "constructive," creative act, which challenges and subverts consumerist paradigms, while simultaneously generating and exploring possibilities for different ways of living in the world (Harkness 2009, p. 155; Hawkins and Muecke 2003).

While drawing on building techniques with a long history, Earthships could be viewed as the brain child of a counter-cultural "starchitect" (Sklair 2010; Sklair and Gherardi 2012). Like many of the 'Earthshippers' (Harkness 2009) interviewed during my research, my interest in Earthships was piqued watching the tale of the charismatic and controversial figure (see Sharpe 2000; Sterbenz 2014) Mike Reynolds in the film *Garbage Warrior* (Hodge 2009). Reynolds graduated from the University of Cincinnati in 1969, his thesis was published in the Architectural Record in 1971, and he built his first house from recycled materials in 1972 (Zimring and Rathje 2012, p. 596). The "Thumb House," used beer cans wired together into 'bricks', which were mortared together and then plastered over, the design for which was patented in the US in 1973 (ibid). *Garbage Warrior* (Hodge 2009) details Reynolds's quest to introduce new legislation and be granted approval to establish a 'test site' in Taos, New Mexico where he could legally experiment with his unconventional building styles. Although reports differ with regards to timing, this experimental approach to building and disputes with some clients led to Reynold's voluntarily surrendering his New Mexico architecture and construction licenses, with most accounts suggesting this happened at some stage during the 1990s (Hodge, 2009, Sharpe 2000, Zimring and Rathje 2012). This also reportedly inspired the renaming of his company from Solar Survival Architecture to Earthship Biotecture, which effectively established a new field of architecture and construction (Open Eye Media 2015). Earthship 'Biotecture' is a portmanteau of biology and architecture and refers to "the profession of designing buildings and environments with consideration for their sustainability" (Open Eye Media 2015).

In 2007 Reynolds' architect's license was reinstated and he has since been invited to lecture at the headquarters of the American Institute of Architects (Zimring and Rathje 2012; Wilcock 2016). Taos, New Mexico is now home to The Greater World Community, a site for training, research, and education for Earthship Biotecture (including the Earthship Academy), as well as an intentional sustainable community, along with other similar communities like STAR (Social Transformation Alternative Republic) and REACH (Rural Earthship Alternative Community Habitat) (Birnbaum and Fox 2014; Freney 2008, 2009; Harkness 2009, p. 65). Reynolds has been commissioned by celebrity Dennis Weaver (Open Eye Media 2015; Wilcock 2016; Reynolds 2008) to build a multi-million dollar Earthship home, a development that was mocked by some participants at Earthship Ironbank as antithetical to Earthship ethics, but is symbolic of the transition of the Earthship into pop-cultural consciousness. Earthships have been featured in their fair share of "architainment" (Sklair 2010), including a host of documentaries and features like: *Earthships New Solutions* (Lozano 2010); an episode of *Grand Designs UK* on the "Brittany Groundhouse," which was also popular among my informants; a follow-up episode (BBC 2009, 2014); Episode Three of Kevin McCloud's (2015) *Escape to the Wild*, which features an Earthship in Belize; and an episode of *Grand Designs New Zealand* (TV3 2015) on Earthship Te Timatanga in the Coromandel Valley. Earthship Ironbank has also featured in popular television in Australia, including an episode on the children's TV show *Totally Wild* (2015) and an episode of *Today Tonight* (2015), focused on 'doomsday preppers'. A number of feature articles have also appeared on Earthship Ironbank to date (e.g. Freney 2006; Husband 2013; Freney 2015; Helbig 2017a, 2017b). Biotecture has clearly spawned a transnational move-ment, with Earthships being built in locations all around the world, including the Asia-Pacific region, both with and without legal approval. There have even been attempts to build Earthship Villages in places like Olst in the Netherlands (Stock-mann 2017) and a proposal to incorporate Earthships into a major city through the Urban Sky Autonomy Project in Manhattan, New York (Earthship Biotecture 2015c).

Earthships are designed and promoted as a kind of "radically sustainable build-ing" (Earthship Biotecture 2015a), which provides comfort conditions year-round, even in challenging climates (Freney, Soebarto, and Williamson 2012, 2013a, 2013b,; Freney 2008, 2009). With their name and operation reflecting the "spaceship earth" metaphor (e.g. Boulding 1966; Fuller 1969), they are described by Reynolds as "independent vessels – to sail on the seas of tomorrow" (1990, cited in Freney 2009). According to Reynolds (2018) Earthships are defined by six main design principles:

1   "Building with natural and repurposed materials" including mud, tires, cans, bottles, and "reclaimed wood and metal";
2   "Thermal/solar heating and cooling" using passive design and enhanc-ing "natural ventilation through buried cooling tubes and operable vent boxes"
3   "Solar and wind electricity" with each containing its own "renewable 'power plant' with photovoltaic panels, batteries, charge controller, and inverter";
4   "Water harvesting" by catching snowmelt and rainwater or *Water from the Sky* (Reynolds 2005);
5   "Contained sewage treatment" with water being used up to "four times, so homes can subsist and even thrive without taking water from the ground or municipal sources";
6   "Food production" involving "interior gray-water botanical cells" and "Aquabotanical systems" which "enhance food production capabilities".

Earthship Ironbank is one of the first council-approved applications of Earthship Biotecture principles in Australia and is also endorsed by Mike Reynolds. This building project used a combination of 'waste' materials (primarily car tires) and 'natural' materials (primarily earth and straw), as well as conventional construc-tion materials, applying Earthship Biotecture principles, to create an Earthship that will function as a living laboratory and bed and breakfast (B&B) accom-modation on Dr. Martin (Marty) Freney's property in Ironbank in the Adelaide Hills. Freney, a lecturer at the University of South Australia, completed a PhD thesis (Freney 2014) exploring the thermal performance of Earthships during the build. His research interests have motivated him to begin to explore the ther-mal performance of Earthships in the Australian environment, with thermal data already being captured at Earthship Ironbank and scope to evaluate the wastewater system, food production, and occupant behavior (energy use, water use, etc.) in future studies.

There is some evidence to support the contention that Earthships can be interpreted as a counter-cultural phenomenon. Schelly (2013, 2017, pp. 124–127) describes Earthship dwellers and builders as part of a coherent sub-culture with its own symbols and features, and Harkness (2009) refers to Earthship builders and dwellers as 'Earthshippers.' My research supported this assertion to an extent, particularly in relation to those participants who had attended the Earthship Academy in Taos and participated in previous builds. The experience of participating in the workshops seemed to provide enough of a sense of "communitas" (Turner 1969) to warrant application of the term 'Earthshippers' (Harkness 2009) to describe participants for the purposes of this paper. However, the workshops at Earthship Ironbank presented more as a confluence of movements, as participants identified with many different overlapping movements/subcultures. For example, many workshop participants shared an interest in permaculture as well as 'alternative', 'natural', or 'eco' building.

Permaculture is an "ecological design system" and "an international movement" (Ferguson and Lovell 2014, p. 252). Australians Bill Mollison and David Holmgren are considered the co-originators of the permaculture concept resulting from their work originating in the late '70s (Mollison and Holmgren 1978; Mollison 1979). According to Holmgren (2002), the three, core ethics, of permaculture are "earth care," "people care," and "fair share (set limits to production and consumption and redistribute surplus)," and permaculture design is informed by the following 12 permaculture design principles:

1   Observe and interact
2   Catch and store energy
3   Obtain a yield
4   Apply self-regulation and accept feedback
5   Use and value renewable resources and services
6   Produce no waste
7   Design from patterns to details
8   Integrate rather than segregate
9   Use small and slow solutions
10   Use and value diversity
11   Use edges and value the marginal
12   Creatively use and respond to change.

The word permaculture originally referred to 'permanent agriculture', due to the initial focus on creating 'permanent' (self-sustaining/sustainable) agricultural systems in response to the rapid spread of high-input energy-intensive industrial agricultural methods (Ferguson and Lovell 2014). However, the term has since 'evolved' to stand for 'permanent culture', as the purview of permaculture has broadened to encompass multiple domains of human existence which are considered to "require transformation to create a sustainable culture" (Holmgren 2002, p. xix). These domains are depicted in the Permaculture Flower (see Figure 11.3), where "Biotechture" (sic) is featured above the "Built Environment" petal as one of the potential solutions to bring about the required transformations (Holmgren 2013, p. 2).

Indeed, Marty is an avid permaculturalist, having completed his Permaculture Design Certificate with his wife Zoe Freney at The Food Forest in South Australia in 1999. Earthship Ironbank was built in a permacultural setting. The Freney family live in a straw-bale shed conversion, with a composting toilet and reed bed greywater system at the top of the block on which Earthship Ironbank was built, growing some of their own fruit and vegetables, harvesting their own honey, caring for the chickens and ducks that run around in the orchard just above the Earthship, and restoring the heritage listed bushland in which the block is situated. The relationship

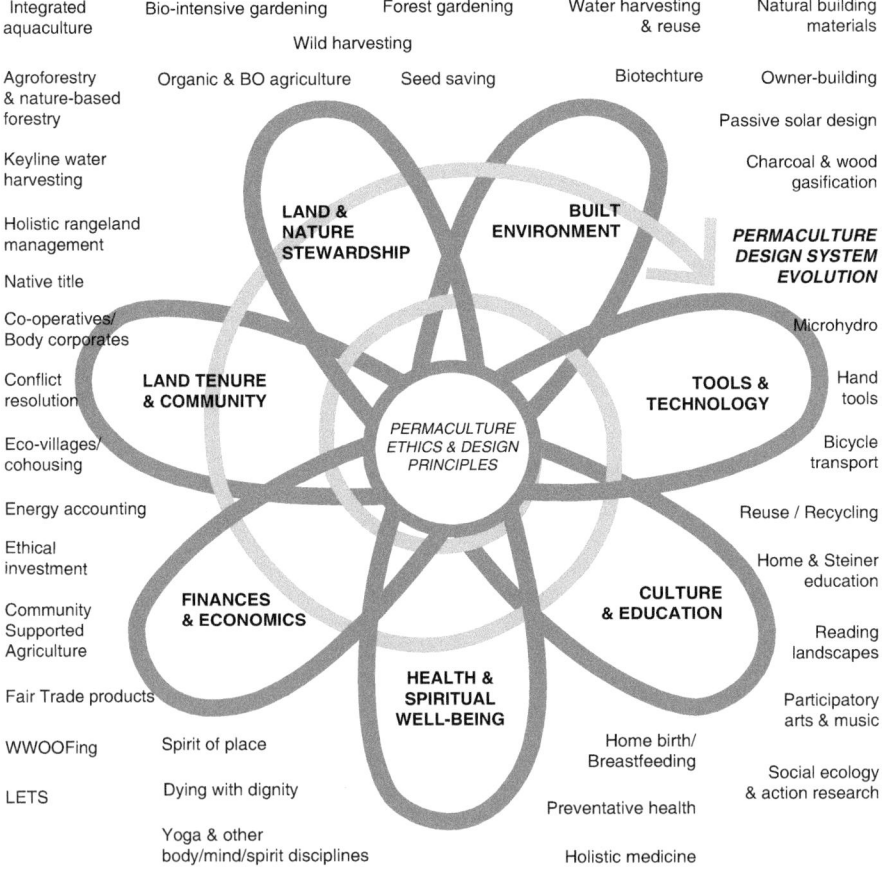

Figure 11.3 The Permaculture Flower

Source: Holmgren (2013, p. 2).

between the Earthship and permaculture movements is also supported by direct quotes from the 'ecological pioneers' involved with originating the movements (Holmgren 2011). In an article from the *Permaculture South Australia Journal* reporting on his experience at the Earthship Academy, Kegan Daly (2014, p. 20) quotes Mike Reynolds as stating, "I think Earthships and permaculture should have sex." In a speech at the Sustainable Living Festival in 2011, Holmgren (2011, p. 2) introduces "one of the grand wizard's (sic) of creative reuse; Mike Reynolds; the Garbage Warrior." While in this speech Holmgren (2011, p. 2) raises some significant concerns about the universal applicability of Earthships in different contexts and climates, he argues "It is the principles behind earthships (and permaculture) that are the universal bits that we must apply, even if some celebrated techniques might not always be appropriate."

Despite the overlaps between Earthship culture and permaculture, it is also important to note that participants were involved in the workshops to varying degrees, and some participants may not have identified with any particular 'subculture' or movement, including permaculture. As Harkness explains (2011, p. 54), Earthship builders and dwellers "often share politics and interests in environmentalism, climate change, and architecture; however, sometimes they do not." As such, for the purposes of this chapter, I will refrain from detailed analysis of how the permaculture principles are reflected in Earthship construction. Nevertheless, of particular relevance to this volume and the Earthship Ironbank workshops, is the focus on waste and reuse common in both movements, as illustrated by the expression "there is no such thing as waste" (FVSS 2013; Harkness 2011, p. 58), reflecting permaculture principle number six, "produce no waste" (Holmgren 2002). While perhaps most apparent in the first of the six Earthship design principles, with one of the most distinctive features of Earthships being their focus on "building with waste" (Hewitt and Telfer 2012, p. 39), the theme of reuse permeates all the Earthships design principles (Reynolds 2018) and was common throughout the workshops at Earthship Ironbank.

Reuse, reclamation, and recycling in earthship construction

Harkness (2009, p. 339) argues that reuse is a common term employed by 'Earthshippers' and found that "people understood the idea of re-use and often praised the Earthship as an alternative to what they saw as 'throw away' culture of today's society." Reuse was viewed as a "necessary task" that connects consumption with its "consequences" (Harkness 2011, p. 61). However, a distinction was drawn between terms like recycling, reuse, and reclamation. For example, reuse or reclamation was typically "associated with a low-energy and less technologically-dependent way of re-deploying materials or resources," whereas "recycling" was associated with "high energy costs and centralised plants" (Harkness 2009, p. 333). While conventional recycling methods can often be quite energy intensive, 'Earthshippers' often reclaim and reuse discarded objects with only slight modifications. However, recycling, reclamation, and reuse are, most likely, all activities carried out by 'Earthshippers' (Harkness 2009, p. 333). For example,

Harkness (2009, p. 333) explains how staff at the Earthship in Fife "both espoused and lived by the mantra of 'Reduce, Reuse, Recycle'," reminding them of "the preferred order of priorities" and the need to avoid consumption where possible, and when required, to buy "second-hand or built to last." They also "maintained, fixed, re-used and re-claimed as much as was possible and they recycled things they could not reuse." This included "recharging batteries," "using wind-up torches and radios," "growing their own vegetables," and "freecycling".

Similar practices were observed at Earthship Ironbank, though a more accurate mantra might be "Refuse, Reduce, Reuse, Repair, Recycle" (FVSS 2013) with many damaged items being repaired with gaffer/duct tape (this is also joked about by Schelly [2017] in her study of Earthship culture), including a phone cover, tools, and other items. Participants often engaged in repair activities during the breaks in workshops, including the repair and reuse of damaged clothing. During both the formal (construction related) and more informal workshop interactions, participants shared information and considered materials in terms of how they might be used, altered, repaired, reclaimed, reused, or repurposed to avoid or minimize overconsumption and its associated waste generation. For example, one evening a crew member explained the conversion of his Toyota Hilux to a "Veggielux," so it could run on waste vegetable oil, along with the processes required to maintain it. During another informal gathering, female participants shared knowledge about the use of menstrual cups and how they had found them better than conventional sanitary products, not only in terms of 'waste' and the health of the environment, but also in terms of their own health, increasing the connection they felt with their own bodies. Other topics that might conventionally be considered 'taboo' (Douglas 1966) were also broached, including the ecological impacts of death and dying and how alternative burial practices could help minimize ecological impacts, reduce waste, and address our disconnection from this process. As such, discourses of reuse were also discourses of reconnection: of consumption and production with their consequences; of people with their bodies and associated "rites of passage" (Van Gennep 1960); with ecology; with material culture; and with one another.

Harkness (2009, p. 332) identifies the extent to which they reuse materials as one of the qualities that distinguishes 'Earthshippers' from other eco-builders. She acknowledges that "re-use is reasonably common in eco-building, as builders strive to use reclaimed timber on their floors or to incorporate other salvaged items into their designs" (Harkness 2009, p. 333). Earthships are no different in this regard as they tend to use "reclaimed timber" and "salvaged masonry" "wherever possible" (Hewitt and Telfer 2012, p. 50). However, Harkness (2009, p. 333) proposes that the difference is that Earthships not only reuse salvaged materials from other building projects, but also "the discards from other spheres of production and consumption." There is a kind of "junk aesthetic" (Hewitt and Telfer 2012, p. 51) to Earthship building that can be viewed as an "act of recovering and transforming the detritus of the industrial age into handmade objects of renewed meaning" (Harkness 2009, p. 334 citing Seriff 1996, p. 9). For example, artistic techniques such as mosaic or earth sculpture are also common in Earthship builds and "salvaged materials are often used in a best-fit sense" (Hewitt and Telfer 2012, p. 51). The Earthship Biotecture (2015a) website states that: "A sustainable home must make use of indigenous materials, those occurring naturally in the local area," which can result in a "House as Assemblage of by-products." In some senses, Earthshippers can be thought of as *bricoleurs*, in Levi-Strauss' (1962) sense of the term, in that they "make do" with materials that come to hand (Harkness 2009, p. 340). However, while there is a strong element of improvisation in Earthship construction, the core planning process and careful selection of materials to meet requirements of the design should not be underemphasized. Rather, Harkness (2009, pp. 9, 40) argues that 'Earthshippers' are more akin to Levi-Strauss' (1962) concept of artists, in that they also act as scientists and engineers, in their engagement in an iterative design process. 'Earthshippers' work with a combination of materials, all of which may not completely reflect the beliefs and values of the builder, who is working not in some illusory utopia, but with the perfectly imperfect matter at hand to meet the practical aim of building something better (Harkness 2009).

Perhaps the most prominent design signature of Earthships is the reframing of used tires as a building material, which is a point of debate both within and outside the Earthship movement. For example, some of the participants in the Earthship Ironbank workshops expressed a preference for other forms of 'natural' or 'eco'-building, specifically because of the concerns associated with use of tires as a building material. Hewitt and Telfer (2012, p. 40) identify these concerns as those relating to "legislative and regulatory difficulties," stemming "from the legal classification of used tyres as waste," along with "concerns about the specific properties of the material itself, regarding leaching, durability, off-gassing and fire risks." However, in his PhD Thesis, Freney (2014, pp. 19–20) argues that:

> While these issues may be problematic in other contexts in which tyres exist, in the context of a tyre wall, they are either not applicable, or they can be managed by engineering or construction methods.

Preston Prinz (2015) provides a summary of research on "tires and off-gassing" that raises potential health concerns in *Hacking the Earthship*. However, the only research cited in this summary that refers directly to the reuse of tires in Earthships, rather than the recycling of tires in other forms like rubberized concrete, is an undated report from Humboldt State University, which states:

> Tires decompose when exposed to high temperatures, sunlight or oxidising agents. None of these elements are present when a tire is packed with soil and surrounded by a stucco barrier inside an Earthship. That being said, the tires used in Earthship walls are of minimal risk to inhabitants because they have little potential to decompose.
>
> (Humboldt State University, n.d.)

Furthermore, the tire bricks used to form tire walls are usually covered over with a final coat of cement, cob, and/or some other form of render, as part of the Earthship building process, further minimizing the exposure of tires (see Figure 11.1). However, the same report concludes:

> EPA researchers have acknowledged that the current literature we have pertaining to the health risks that used tires pose is incomplete, and that further study should be conducted before used tires are used in applications where humans are exposed.
>
> (Humboldt State University, n.d.)

A research report commissioned by Ecoflex Australia (2005, p. 1), a company that builds with waste tires in Australia, states that "the use of tyre derived products for civil engineering applications can be environmentally safe, beneficial and superior to alternative recycling strategies." However, further independent research into their application in dwelling places is clearly warranted (Harkness 2009).

As Harkness (2009, p. 316) points out, much of the controversy around the use of tires and other materials in Earthships, results from the transgression or transcendence of boundaries, for example, between understandings of the 'environmentally-friendly' and 'natural', and that of the 'industrially manufactured' and 'man-made', along with the taboos associated with trash (Douglas 1966). Despite this controversy, tires continue to be a prominent material in Earthship construction. One obvious reason for this is the sheer scale and rapidity at which tire waste is currently generated, along with the social and environmental costs associated with the ways in which this 'waste' is predominantly dealt with. According to the Australian Government (2014), of the approximately 48 million tires that reached their end of life in Australia in 2009–2010, "only 16 percent were domestically recycled", "18 percent were exported", and the remaining "66 percent" were either disposed to a "landfill", "stockpiled", "illegally dumped," or categorized as "unknown". Similarly, the report from Humboldt State University referred to above (citing an old EPA report), states that every year "Americans discard approximately 290 million tires" or "roughly one tire per U.S. citizen," of which 20 percent end up in landfills and 80 percent "are either combusted as an industrial fuel source, or are used in civil applications such as rubberized asphalt roads" (n.d.).

'Earthshippers' who build with tires argue that reusing tires in an Earthship is preferable to either having them end up in a landfill or reusing them in many of the dominant current reuse applications for tires, both of which can have significant social and environmental impacts (Harkness 2009, p. 31; Hewitt and Telfer 2012, p. 40). For example, the Humboldt State University (n.d.) report states that "about 130 million tires are incinerated each year for industrial fuel uses," resulting in "dioxin and furan emissions, which are highly toxic and dangerous for biological health; not to mention the onslaught of CO_2 emissions, which cascade global warming." As Hewitt and Telfer (2012, p. 51) argue, the benefits of the reuse of tires in Earthship construction are twofold in that the process "makes use of the man-made materials that are so readily available, short-circuiting the manufacturing process and reducing the embodied energy" as well as negating "the need for direct harvesting of virgin natural materials such as timber, by using reclaimed materials instead." The 'tire brick' is also composed predominantly of earth, a substance that is, arguably, one of the most 'sustainable' building materials available (Sheen 2009; Hickson 2015).

There are also practical and operational reasons for the use of tires in Earthship construction. Reynolds (Telfer 2003) once claimed, "If I was paid $30 million to invent the best thermal mass brick I could, I would invent a tyre." As Reynolds (1990, p. 78) has stated, in an example of permaculture's suggestion that the "problem is the solution" (Mollison 1988, p. 15), "The very qualities of tyres that makes them a problem to society (the fact they won't go away) makes them an ideal durable building material for Earthships." Earth-rammed tires provide excellent thermal mass, taking somewhere between 75kg according to Hewitt and Telfer (2012, p. 42) and around 300 pounds or 136kg, according to Reynolds (1990, p. 83), of earth per tire. The tires are usually first lined with cardboard, then shoveled full of earth, before being 'rammed' using a sledge-hammer. The resulting 'tire bricks' are stacked to create walls that are earth rammed and around "three to four times thicker than conventional walls" (Hewitt and Telfer 2012, p. 42). Once used in Earthships, tires have relatively low embodied energy as they do not need to be manufactured to be used in the buildings, merely rescued from landfill. They can be pounded by hand within around half an hour by an enthusiastic worker with a sledgehammer. The round shape of these 'Earthship bricks' also lends itself to the formation of more 'organic' structures, which were often admired by workshop participants, both at the Ironbank site and the sites of other alternative, natural, or eco builds visited during the workshops, where participants frequently remarked on the 'sexy-curves' of the buildings and took time to hug and stroke them. The main 'tire walls' at Earthship Ironbank are indicated in Figure 11.2 in the thick black line on the top of the building floor plan, which curves around into the white line, indicating the tire wall that forms the entrance (see rendered over tire in entrance picture in Figure 11.4). Tires can also be used in landscaping, as they were at Earthship Ironbank, for the formation of steps, ponds, and other features, where they may or may not remain partially exposed (see Figure 11.1).

However, tires not only "perform valuable structural and thermal functions" in Earthships, but "also provide an implicit critique of attitudes to waste: if a material can still be useful, then shouldn't it be used?" (Hewitt and Telfer 2012, p. 39). Given the association of many contemporary ills with the accelerations and increasing 'mobilities' (Urry 2007) afforded by contemporary forms of corporate consumer capitalism, there is a powerful symbolism associated with ramming a tire, formerly an instrument of mobility driven by fossil fuels, full of earth to 'root it in place'. In Earthships the tire is transformed "from an index of a wasteful and automobile-crazed society to the crux of a building movement with environmental and social concerns" (Harkness 2009, p. 331; 2011, p. 56).

Earthships transgress boundaries, not only in their building with waste materials, but also in their transgression of boundaries between the 'natural' and 'environmentally-friendly', and the 'industrially manufactured' or 'man-made', in materials selection (see Harkness 2009). For example, Reynolds (2018) refers to tires as "a globally available 'natural resource'." Furthermore, one of the concerns raised by Preston Prinz (2015, p. 53) in her summary of research findings on Earthships, is that they use "many more manufactured materials than not" and are "therefore not as sustainable or natural as most people believe." As Hewitt and Telfer (2012, p. 51) argue, "Inevitably, the greater part of the specification involves newly manufactured products that are needed to make the earthship the high-performance structure that it aspires to be." For example, such materials, many of which are also employed on conventional builds, include rainwater tanks, PV panels, battery systems, plastic as vapor barriers, various forms of lining, gravel, plumbing fittings, wire, rebar, and, as mentioned earlier, often a fair amount of duct tape (e.g. see Figure 11.8).

In particular, the amount and usage of cement in Earthship construction has attracted criticism (Hewitt and Telfer 2012, p. 49). However, the use of cement is "seen as unavoidable" by some, allowing Earthships to withstand the elements, and providing a sense of permanence and assurance of "'Lasting' in Time" (Harkness 2009, p. 328). Like most Earthships, Earthship Ironbank did involve the use of a substantial amount of concrete, and for the most part, Portland cement was used. However, concerted efforts were made to reduce the amount of concrete utilized in construction at Earthship Ironbank. For example, the tire walls were packed out with cob, there was experimentation with Hempcrete as an insulating material over the vault, and Earthship Ironbank also featured an earth floor, which reduced the amount of cement used when compared with, for example, a concrete slab floor common on conventional builds. Although Earthships often feature a concrete beam on top of the tire wall, the concrete beam can be smaller than a wall footing, and there were no concrete footings under the tire walls at Earthship Ironbank. Harkness (2009, p. 332) contends that the use of cement does not occur without serious consideration of "the past history of cement (and the future for both themselves and their environment) in their avoidance of it" where possible. For example, while cement mortar is often employed to 'glue' the bottles and cans together, "infilling with various types of waste, such as bottles is used in an attempt to reduce the amount of cement being used overall", rather than "packing out the spaces between tyres solely with cement" (Hewitt and Telfer 2012, p. 49). This is particularly the case with non-loadbearing walls, which "offer the opportunity to experiment with different types of 'brick' that might otherwise merely be discarded into landfill", including bottles (usually glass, but I have seen plastic used as well – though not at Earthship Ironbank), and aluminum cans (see Figure 11.4) (Hewitt and Telfer 2012, p. 49).

One of the most visually impressive features of Earthships is the glass bottle wall features that are often created from the patterned embedding of 'bottle bricks' into the cob or cement walls (see Figure 11.4). These bricks are formed by using a cutting tool to cut the necks from two bottles and sticking the remaining bases together with tape to create "a uniform cylinder" that can be used to form non load-bearing feature walls and has significantly reduced embodied energy compared with conventional bricks designed for the same purpose (Hewitt and Telfer 2012, pp. 49–50). At Earthship Ironbank these 'bricks' were created at a special 'bottle brick station', where bottles were systematically selected, washed,

Figure 11.4 Entrance vault to Earthship Ironbank featuring can and bottle walls

Photo: Keri Chiveralls.

and individually cut with a circular diamond saw, the bases taped together with gaffer tape, and the necks smashed into a large green bin. The shards in this bin were later used to create a stunning patterned floor by adding them to the mix for the bathroom floor, which, in the right light, sparkles with the tiny different colored shards. There was a casual art to selecting the right bottoms to match together. For example, it was not always preferable to match two of the same colored bottles together to form a brick as the darker colored bottles often worked best when matched with a clear base to enable better penetration of light (see Figure 11.5). As Hewitt and Telfer (2012, pp. 49–50) argue, the "aesthetic beauty" that this reuse creates and the "interplay of shape and color" as the light permeates through the bottles demonstrate "the amazing capacity of rubbish to be turned into something inspiring" (for more on reuse of bottles, see Harkness 2009, p. 311).

However, as mentioned above, bottles are not the only beverage container used to create signature wall features. Demonstrating the broad definition of "natural

Figure 11.5 Bottle brick at Earthship Ironbank

Photo: Keri Chiveralls.

and indigenous materials" employed by 'Earthshippers' (Harkness, 2009), Reynolds (1990, p. 79) describes "the aluminium beverage can" as "a little durable aluminium brick that appears 'naturally' on this planet" and is "indigenous to most parts of the planet that are heavily populated." Such cans are often used "to create eye-catching features" in Earthships, "such as panel walls, domes and vaults" (Hewitt and Telfer 2012, p. 49, and for more on reuse of cans, see Harkness 2009, p. 311). However, as mentioned above, the reuse of materials in Earthship construction is questioned by those considering the benefits from a material life-cycle perspective. For example, Hewitt and Telfer (2012, p. 49) argue that:

> non-ferrous metals have the highest embodied energy of any material that is used in volume in any industry or sector in society, and it is better that they are recycled conventionally in order to reduce the need for raw material extraction. This is particularly the case when the aluminium can is not a functional element – that is, it provides no structural or thermal quality to the building.

From a material lifecycle perspective, this practice could be argued to make little sense in a state like South Australia, which was the first Australian state to introduce a container deposit scheme and has a well-established recycling infrastructure for aluminum cans and bottles (EPA 2016). However, as Hewitt and Telfer (2012, p. 49) point out, "This technique evolved in the 1970s in the desert, where there was no recycling infrastructure." In the future, with better recycling methods, these cans (and bottles) may be able to be recycled, even after use in an Earthship, but for the functional life of the building they remain "out of the loop."

While the material lifecycle implications of this kind of reuse may be problematic, Harkness (2009, p. 332) argues that 'Earthshippers' are aware of "the material's 'life story'." The "life story" of the cans and bottles used to "pack out" the walls of the Earthship as well as form decorative features, was of particular significance at Earthship Ironbank, as much of the contents were consumed on site, or containers were carefully selected and held over from other social gatherings offsite. In a similar process to that described by Langford (this volume), 'Earthshippers' (Harkness 2009) could be argued to have developed a form of community-specific "skilled vision" (Grasseni 2007), or "Earthship vision," which enables them to see Earthship potential in different objects. For example, on the trips to local tire stores, participants learned to look for tires with specifications suited to the different construction tasks being completed. Similarly, drinks were often selected and purchased with consideration of the suitability of their containers for the patterns being developed for the walls. For example, the blue and green bottles were particularly prized, as was the gin bottle with a square base that was used to create a diamond-like shape in the feature walls. As such, building with bottles and cans enabled the building process to extend into the social gatherings that took place after building had ceased for the day, and ensured that some of the material generated by these activities formed part of the Earthship body. The relationships formed throughout the building process were reinforced through this process of selecting materials and building with waste, symbolically embedding the social body of the workshops in the physical body of the building, creating a sense of permanence from impermanence, and positioning the transient experience of the Earthship Ironbank workshops within the context of broader social and physical structures. As such, in accordance with Haraway's (1997) "cyborg theory" and Alaimo's (2010) concept of "transcorporeality," Earthships also challenge traditional subject/object divisions, emphasizing the permeability and fluidity of boundaries and the connection between builders and buildings.

Reuse in Earthships could be argued to emphasize "energies of social remembering and disclosure" over techniques of "social forgetting and veiling" (Appelgren and Bohlin 2015 p. 156). There is a certain "wizardry" in the way in which the violence (symbolic and actual, social and environmental) inherent in the relations of production and consumption associated with corporate consumer capitalism are often rendered invisible, as described by Crocker in the opening chapter of this volume. Earthships employ a different kind of "wizardry" to reorder such relations and render waste visible as part of a broader process of "making things visible" (Harkness 2009, p. 355). While there is usually much literal rendering employed, effectively making most of the tires utilized in Earthship construction invisible, from the year that the first tire was pounded on site, to time of writing, there have been a large number of tires that remain exposed and visible at Earthship Ironbank. Even on completion of construction, eco/alternative/natural buildings often employ 'truth windows,' and in Earthships 'waste' practically becomes a window in the glass bottle walls. Earthship Ironbank also features a 'truth window' on the main tire wall. Schelly (2014, p. 679–680) explains the importance of truth windows as a form of 'claim making' in the "world of alternative residential construction":

> when homes are constructed out of alternative materials such as straw or tires and then covered with earthen plaster, homeowners will often include 'truth windows' – openings that allow them to see the alternative building form underneath the finished exterior – so that they can continually open the door and see their claim to a world that is now built around them.

Scanlan (2005, p. 121; see Harkness 2009, p. 293) asserts that due to the levels of "specialist production and public bureaucracy" in industrialized societies, people are "already one step removed from the consequences" of their own waste and "need never see it". In her discussion of the pipework and wiring in Earthships, Harkness (2009, p. 116) states that Earthships don't "break too many of the taboos about what should be hidden and what should be on show." However, the visibility of 'waste' in the form of truth windows, bottle and can walls directly and deliberately challenges such taboos.

Another significant component of reuse in Earthship construction is the reuse of earth. While tire-pounding is seen as the quintessential Earthship activity, this is just one of the variety of ways in which earth is reused in Earthships (Harkness 2009, pp. 31, 321). Earthships are usually backed into a large mound of earth or 'berm' that provides thermal mass and ensures Earthships are literally and metaphorically "part of the earth" (Harkness 2009, p. 296). Substantive earthworks are often required to 'lay the groundwork' for Earthship construction and much of the earth obtained through this process that is not used to construct the earth berm, is reused in tire-pounding or the cob mixes created through collective 'cob stomping' (see Figure 11.6), and used at various stages throughout the building process. Much of the earth that is 'reused' is earth from onsite. However, where additional earth is required for construction, 'Earthshippers' (Harkness 2009) will try and seek earth for reuse from other construction projects, as was evident in

Figure 11.6 Cob stomping at Earthship Ironbank
Photo: Keri Chiveralls.

the "clean fill wanted" sign often displayed at the entry to Earthship Ironbank. The connection between builder, the building, and the environment is reinforced through such earth building techniques. For example, the high-energy input and 'sweat equity' required from techniques like tire pounding, and the literal bodily engagement with the materials that become part of the structure of the Earthship, required for example in cob stomping and application (which often took place directly with bare hands and feet), reinforce the sense that the builder is 'leaving their mark on' or 'putting themselves into' and becoming part of the building and its life story. For example, at Earthship Ironbank, even the dogs brought onsite by the workshop participants became part of the building, with their paw prints left behind in the cement on the tire steps, as well as in the earth (cob) floor.

Earthships challenge dominant cultural constructions, taboos, and understandings of 'waste', treating substances often associated with 'pollution and taboo' as valuable resources and reinforcing the idea that waste is simply "matter out of place" (Douglas 1966). For example, conventional Earthships use water "four times" (Reynolds 2018): firstly for general uses like drinking, bathing and laundry; secondly, the greenhouses which also support the passive solar design, contain a planter bed and water treatment system that reuses "waste" greywater through planter treatment cells, planted with a variety of plants selected for wastewater cleaning properties, along with beauty and food production (e.g. the banana palm in Figure 11.7 which can do all three); thirdly, this water is usually then directed to toilets for flushing; and finally water is directed to an outside septic tank which overflows into an outdoor landscaping "botanical cell". Following the logic that those elements that most challenge dominant understandings of 'waste' will be associated with the most 'taboo', the wastewater system was the system that caused the most difficulty in terms of legal regulations and requirements at Earthship Ironbank, with SA Health requiring that elements of the wastewater system be removed to ensure the Earthship complied with legal requirements (see Figure 11.7). However, if anything was lost in terms of food production from the greenhouse, it appears to be well compensated for by the food production taking place on the berm, with pumpkins and watermelons cascading down the hillside at time of writing. Along with challenging taboos, norms, and regulations, the movement of water through the Earthship decries waste "as a false category" (Harkness 2009, p. 301). For the 'Earthshipper', 'waste' becomes fluid, subject to reflexive consideration and rendered visible again; "matter is always flowing but it is never 'away'" (Harkness 2009, p. 301). As such, Earthships challenge the boundaries "drawn between waste and non-waste in their wider societies" demonstrating how "boundaries can be re-drawn" and providing "an important counter-measure to the escalation of 'waste' in the wider society" (Harkness 2009, p. 306).

Earthships also challenge dominant approaches to waste and consumption through their passive design, 'off-grid' energy, water, and wastewater systems that encourage frugal use of these resources and encourage consideration of the impacts of products that enter the system (for example cleaning products must suit requirements of the septic and water treatment systems). Schelly (2014, p. 677) explains how the "Earthship ethic" is "an ethic of justice" informed by an emphasis on responsibility and self-reliance afforded by the "off-grid nature" of Earthships:

> People who live in Earthships talk about their homes as living organisms;
> their homes are designed to provide all of their water, energy, and heating and

Figure 11.7 Planter at Earthship Ironbank with banana palms
Photo: Martin Freney.[1]

cooling needs without input from any mainstream technological infrastructure, yet this requires that homeowners interact with their homes by opening skylights or shutting blinds to maintain thermal comfort, changing the filters that provide clean water and allow them to treat their own waste water, and monitoring the battery storage systems that provide electricity.

This also requires that Earthship dwellers maintain an awareness of their environment (including seasonal and weather variations) and monitor and adjust their consumption patterns to keep within the limits, for example, of the energy stored and water captured in the Earthship. Similarly, Harkness (2009) describes how 'Earthshippers' embody the "land ethic" espoused by Aldo Leopold (1949). Harkness (2009, p. 307) and proposes that an 'Earthshipper' displays a logic similar to the thinking of the "organic gardener or permaculturalist". However, the "Earthship ethic" (if there is one, and I would argue, that there are many) is not limited to either the land ethic of "earth care" (Holmgren 2002), or the "fair share" ethic of "social justice" (Schelly 2017, pp. 145–148), which sets limits on production and consumption (Holmgren 2002), sees constraints as positive, and "turns away" from "pursuit of things" purchased and "the perpetual growth of needs" (Harkness 2009, p. 119). Rather, Earthships could be seen to reflect all three permaculture ethics as they also emphasize "people care" (Holmgren 2002), in the sense of being designed to look after people as people look after them, and, at least in the case of Earthship Ironbank, of co-creating relationships and social structures along with the building of physical structures.

The tensions between independence and interdependence, also reflected in the broader permaculture movement (e.g. see Holmgren 2017), with the dual focus on disconnecting from wider society by going 'off-grid' and working together to create a better model, are partially resolved in the Earthship movement through a logic focused on breaking down dichotomies, which recognizes that "Earthship autonomy is reliant upon interdependence" (Harkness 2009, p. 120). The off-grid focus of Earthships, to some extent, succeeds in cutting Earthshippers off from complex global systems, but "also acknowledges the connections or relationships that our being-in-the-world creates" (Harkness 2009, p. 119). Harkness (2009, p. 122) argues that the term "dwelling" in relation to Earthships refers not only to the process of inhabiting a building, but also to a different way of inhabiting the world, it is a way of life that "provides an example of an alternative form of dwelling, and so of building and being-in-the-world."

DIY, DIT, and collaborative prosumption: remembering the past and shaping the future?

Many of the counter-hegemonic elements evident in Earthship culture are not unique to Earthships, but rather draw from and reflect long-standing counter-cultural traditions. For example, Harkness (2009, p. 203) points out that while "Earthships are a phenomenon of the late twentieth and early twenty-first century", they aim to solve problems that are similar to those tackled by the Arts and Crafts movement in the latter half of the nineteenth century. This movement "was born of reaction to industrial society's 'cheap mass-produced goods'" and rebellion against the complacency of the middle class who seemed content to ignore the negative impacts of industrialization, mass production, and factory work (Harkness 2009, p. 203, citing Tinniswood 1999, p. 6). Accordingly, as can be seen in various examples of a contemporary resurgence of such movements (see Luckman 2013, 2015), there is an emphasis on craftsmanship, the "low-tech" or "slow-tech" (Price 2009), and the "hand-made" in Earthships, as can be seen in the intricate sculptures formed of cob on the inner walls of the greenhouse at Earthship Ironbank (e.g. see Figure 11.8). Earthshippers emphasize building as a craft, rather than an industry (Harkness 2009, p. 189), sacrificing (Maniates and Meyer 2010) the time benefits associated with industrialization and mass-production, for the gains of spontaneous problem solving, interactive and bespoke design, and creative expression.

Ritzer (2014a) contends that while production and consumption have always been intertwined, binary understandings of production and consumption as hegemonic concepts have obscured and obfuscated understandings of the expansion of new forms of 'prosumption' (Toffler 1980) (activities involving both production and consumption) so dramatic as to herald the "age of the 'prosumer.'" Earthship Ironbank can be seen to reflect emergent shifts not only to "collaborative consumption" (Botsman and Rogers 2010), but also to "collaborative production" or "prosumption" (Ritzer 2010, Ritzer and Jurgensen 2012), embodying the ethic not only of do-it-yourself (DIY), but also of do-it-together (DIT). The process of tire-pounding is admittedly very labor intensive and somewhat time consuming, as is the entire process of Earthship construction. However, this process has led to Earthship construction being associated with large collaborative building workshops and "work parties" (Harkness 2009, p. 205), which address the intensive labor requirements of the process while offering countless intangible benefits to participants (and can be a great deal of fun). Whether deliberate, preferred, or otherwise, this process presents a challenge to the acceleration and time pressures associated with industrialized rapid production.

The Earthship Ironbank workshops consisted of a paid crew, with varying levels of experience, many of whom had trained at the Earthship Academy in Taos, and volunteers with a wide range of skills and experience, many of whom had paid to attend the workshops and to learn new skills while sharing in the experience of Earthship building. The money paid by attendees to participate was used to contribute to covering the costs of the paid crew, shared food, and some materials. The payment structures for such workshops is a point of controversy for the Earthship movement. As discussed by Harkness (2009, p. 197), moves by Biotecture "to charge interns for the experience of Earthship building was regarded with particular caution" by some, and the structure of the Earthship Ironbank workshops provides "further evidence of a shift back to measured exchanges of time, effort and skill (albeit a set of still increasing skills) for money, with some self-builders charging volunteers to learn on the job." However, like the volunteer labor discussed by Harkness (2009, p. 196), there was an acknowledgement at Earthship Ironbank that "labouring should also be an educational and enjoyable

Figure 11.8 Tree sculpture on inside wall of greenhouse at Earthship Ironbank
Photo: Keri Chiveralls.

experience for the volunteer." The workshops involved similar "subtle social mechanisms" that helped regulate the nature of these works divisions, including the provision of healthy and delicious food for all, and an understanding that the workshops needed to be a pleasurable, fun, and rewarding learning experienced for all involved (Harkness 2009, p. 196). Labor was not to be "taken for granted" and people, especially those who weren't paid crew, were free to participate in, or not participate in, tasks as they wished (ibid). However, there were certain shared social obligations associated with communal living on site, that all were expected to meet. As such, the workshops reflected the phenomenon of "playbor" (combining labor and play), described by Ritzer (2014b) as having much in common with "prosumption".

Despite the differences in economic investment between those who paid to attend, those who were paid to attend, and those who were permitted to attend and volunteer for short stints, in practice differentiation between crew and volunteers was often fluid, and lines between the two frequently blurred. For example, once a beginner had successfully pounded a tire or two, they would often become the 'expert', being turned to by other beginners and taking on the role of instructing and assisting newcomers to the workshops. Crew encouraged volunteers to participate in a self-directed manner. Marty and/or crew would outline the "designs and dreams," visions and aims for the day, and once the general tasks had been explained, it was trusted that participants would be able to observe, learn, and determine what needed doing for themselves (Harkness 2009, p. 122). There was no culture of "spoon-feeding," with the onus "on the learner to ask questions, and sometimes even to ask for new tasks" (Harkness 2009, p. 184). Such findings support Harkness's (2009, p. 185) contentions about the value accorded to "learning by doing" and "*being there*." This was reflected to some extent at Earthship Iron-bank in the level of status afforded to those who had 'logged more time', whether through the academy, previous builds, on-site, or relevant life experience gained elsewhere.

However, innovation was also recognized and celebrated, along with experience. While important decisions were often made by crew in consultation with the owner, for the most part, solutions to problems were discussed openly with input often sought from those who weren't officially designated as crew, particularly those with the most knowledge or experience in the relevant aspect of construction, and input from all was welcome. At Earthship Ironbank, individual ingenuity in approaching tasks was encouraged and the acknowledgement that there were many different ways of approaching a task was reflected in the catch phrase from the workshops, often given in response by a particular crew member when asked how a task should be performed, of 'Choose Your Own Adventure.' This phrase also featured on the Ecoshout (2015) website in an advertisement for an Earthship Biotecture build in New Mexico and was repeated by many at the workshops at Earthship Ironbank.

In her discussion of 'craftivism,' Luckman (2013, p. 260) argues for the re-imagining of "a more fluid spectrum of 'pro-am' activity," acknowledging the overlap between the categories of professional and amateur. Similarly, Harkness (2009, p. 119) outlines the importance of non-expert knowledge and cultural exchange in Earthship building. The workshops at Ironbank were permeated by an atmosphere of mutual learning, co-operation, co-creation, and a sense that everyone has something to contribute. Learning, work, and play activities all had the potential to spontaneously adapt and evolve based on input from participants who had something to contribute, regardless of status. This DIT ethic of the workshops was reflected in the conviviality, social interaction, and cultural exchange that took place throughout: from sharing meals, working side-by-side, and often sleeping side-by-side, through the quiet moments of reflection and discussion in the shade, to the creative explosions of celebration like the 'Frock up Friday' dress-up parties that occasionally accompanied 'beer o'clock' drinks at the end of the workday. The workshops allowed for the collective design and structure of work and play, whether this took the form of the more formal impromptu presentations on particular areas of knowledge relevant to Earthship building; the informal discussions and sharing of information that took place while working side-by-side; the spontaneous songs or performances; or the sharing of tools, instruments, and costumes people had brought with them. Such interactions facilitated learning and information exchange across many of the permaculture domains depicted in Figure 11.3, along with the development of social networks and connections that continued long after the workshops were over. This is also evident in the organization and proliferation of subsequent workshops, and even businesses, that were launched in the wake of the workshops at Earthship Ironbank, along with the connections on social media in Facebook groups, events, and other forums.

As Crocker has argued (this volume, referring to Rosa 2003), acceleration in modernity is typically accompanied by a countervailing desire for various forms of deceleration. A deliberate attachment to the 'slow' and 'old' is often adopted, as a form of resistance in the face of a surrounding and more dominant culture of acceleration (ibid). Compared with dominant approaches to contemporary construction, Earthship building could be considered a "slow solution" (Holmgren 2002): a form of 'slow building' or 'slow architecture'. The first tire was pounded at Earthship Ironbank in 2009, and the doors were first opened for business as a B&B in December 2016. Crocker (this volume) describes the process of desynchronization as a drive to escape acceleration through a 'return to the past'. While between one third and one half of the world's population still live in homes built from earth, the focus on building with earth in Earthships could be seen to represent a return to the 'old ways' (Sheen 2009), as earth has been used as a building material for over 10,000 years (Hickson 2015). As argued by Chau in this volume, building with 'waste' or 'spolia', also has a long cultural history. Even in Roman times builders employed the technique of 'cracking off' and incorporated bottles and other 'used' containers as light features in adobe structures (Earthship Biotecture 2015b).

Earthships and permaculture both originated in the late '70s, following the rise of the 'counterculture' of the '60s. In Chapter 1 (this volume) Crocker argues that this 'counterculture' tended to draw on an imagined past to support its 'back to nature', communitarian ethos, and fascination with DIT culture. Earthship building and permaculture are also both affiliated with the "appropriate technology movement" (see Schelly 2013, 2017, pp. 128–131) and *The Whole Earth Catalog* (Brand 1968–1998), which Crocker associates with this "return to the past." Just as the 'back to the land' movements of the '60s were often linked with dystopian fears like nuclear holocaust, as evident in *Today Tonight*'s (2015) feature on Earthship Ironbank, Earthships and permaculture are often associated with dystopian cultural narratives, from those concerned about the impacts of climate change and resource depletion to the 'zombie apocalypse', and 'cults preparing for the collapse of civilization' mentioned in the story. Permaculture in particular emphasizes the need to prepare for various "future scenarios" of "energy decent" (Holmgren 2009).

However, I argue that the processes of deceleration reflected in Earthship culture and permaculture are not about 'returning to the past', but rather about 'remembering' and 'reclaiming' elements of the past to build a different future. Earthships are also inherently utopian, as evident in Reynolds' vision of "independent vessels to sail on the seas of tomorrow" (Reynolds 1990, Introduction; Harkness 2009, p. 122). As Harkness (2009, pp. 348, 121–122 citing Merrifield 2000, p. 173), suggests, this utopian quality should not be dismissed as fanciful dreaming; rather, Earthships turn "dreams into reality" though a process of cooperative imagination and action to co-create a grounded, practical, "concrete utopia" embedded in the earth. Rather than opposing modernity or tradition, "it is more plausible to argue for the convergence of streams of history at the site of the Earthship" as "there is no stark divide between the traditional and the modern" in Earthship building (Harkness 2009, p. 337, p. 242).

This is best illustrated through the recounting of an incident at Earthship Ironbank. Towards the end of one of the many long, hot days of the workshops, a few of us were relaxing on the roof of the Earthship while some of the crew and volunteers continued to work, finishing up tasks they were hoping to complete or make progress on before the light was lost. Our relaxation was momentarily interrupted by frustrated exclamations from below of 'socket-suckers' and 'technohippies', from one of the elder participants in the workshop who had been trying to find a place to plug in his power tools and had encountered the array of mobile phones that were lined up to charge. This may not have been the first time that term 'technohippies' had been used, and it certainly wasn't the last. Regardless, something about this term clearly resonated with participants as it was used again and again in online and other conversations, even years after the event. For example, when a couple who had met at the workshops had a child together, the child was made a 'onesie', hand-painted with the word "technohippy." The term, which had been leveled both lovingly and accusingly at the group on the rooftop, reflects the existence of both a "nostalgia for the past" and "nostalgia for the future" (Bey 2011, p. 68) within the Earthship movement and affiliated moments like permaculture, akin to "reflective nostalgia" (Boym 2001; and see Dawdy 2010 and Crocker, this volume).

The concept of "reflective nostalgia" (Boym 2001 and Crocker, Chapter 1) acknowledges that we cannot return to the past, but can reflect on the past and engage with and adapt elements of it in the present to create narratives about, and practical projects for, the future. This Janus-like nature of these movements that look not just nostalgically, but hopefully and simultaneously, backwards and forwards, like the mythic Roman God, is reflected in references to permaculture as "the cutting edge of a ten-thousand-year-old technology" (Dolman cited in Birnbaum and Fox 2014, p. 3). Permaculturalists have described themselves as "cultural pioneers [. . .] adapting for the future with a connection to ancient principles" (Birnbaum and Fox 2014, p. 3). They have argued that "moving forward in a healthy direction means a return to the ethics that sustained healthy cultures of the past", although this necessarily brings with it an acknowledged need to be wary of "perma-colonialism" and the appropriation, romanticization, and essentializing of other cultures (Birnbaum and

Fox 2014, ibid, p. 10). Similarly, Earthship construction can be best thought of as a reinterpretation of the past that is at once present and future focused, and encourages participants to "Think Long-Term; Act Now" (Harkness 2009, p. 263).

This process of 'remembering' in Earthship building refers not only to elements of "reflective nostalgia" (Boym 2001), but a return to embodiment and working with the body (see Harkness 2009, p. 201 and Smith, this volume), as evident in the significance of human labor and working with hands and the way in which group 'membership' is reinforced through action and physical presence. The term also refers to the counter-hegemonic aspect of Earthship construction in that Harkness (2011, p. 61), citing an Earthship website tagline, argues that Earthships seek to "put housing back into the hands of the people". Additionally, the term reflects the value and significance of the memories held in objects (see Butler and Liboiron, this volume). For example, Harkness (2009, p. 201) describes how for 'Earthshippers', "the relationships and meanings embodied in the object or building which is hand-built are considered as not only more perceptible, but as both more valuable than those of machine manufacture and as mutually reinforcing or complementary." However, 'Earthshippers' are "pragmatic" and part of the wider society they come from; they do not eschew tools and technology; they "are not Luddites" (Harkness 2009, p. 206). Earthships and permaculture cannot be reduced solely to anarchoprimitivism, as participants are very much open to the use of human-scale appropriate technology to facilitate the creation of future infrastructure (Smith in press). For example, Harkness (2009, p. 207) describes how, in her interviews with 'Earthshippers' about the labor involved in tire-pounding, concerns were raised about the introduction of technologies that may assist with speeding up this process. However, some of the 'Earthshippers' from Earthship Ironbank have since taken to using a pneumatic rammer to speed up the process on recent builds and have even established their own company, the logo for which features a pneumatic rammer, promoting this innovation.

This should not be so surprising, as 'Earthshippers' have historically tended to openly embrace technology when it is seen to serve ecological purposes; for example in the use of 'alternative technology' like renewables, computers, and mobile phones (Harkness 2009, pp. 206–208). However, the emphasis is on appropriate technology and the use of "convivial," "*limited* tools" rather than "escalation", in the context "of reciprocal socio-environmental relations" (Harkness 2009, pp. 215, 218, 219). Such concepts are reflected in the permaculture movement through the focus on deliberate descent, degrowth, and voluntary simplicity or "living simply so that others might simply live" (Alexander and Garrett 2017). These movements are united by an understanding that the "disembedding of technical from social relations is associated with a promethean logic of escalation (which might, in neoliberal terms, also be called 'growth' or 'development'")" (Harkness 2009, p. 215). Alternately, these movements seek to "socially re-embed or reinstate the technical move in a different direction" (Harkness 2009, p. 215). Such movements are grounded in an understanding that people need to not only obtain or consume things, they need also to make or produce things and in doing so have freedom to express their tastes, shape their identities, and to care for themselves and others (Illich 1973, p. xxv, cited in Harkness 2009, p. 219). While it could be argued that many paid to attend and 'consume' the experience of the Earthship Ironbank workshops, the workshops served to create an "alternative and thriving system of value" where "value" is emergent from action and "human beings" are conceptualized as "creators" rather than consumers (Harkness 2009, p. 220). Thus, Earthship building can be viewed as a form of "practical-critical" action or praxis, "A Project of 'Radical Dis-Alienation'" which challenges conventions relating to processes of production and consumption, including economic and labor practices (Harkness 2009, p. 21, pp. 16 citing Marx 1977, 156). However, Earthship building can also be seen as a form of "active hope" (Macy and Johnstone 2012), or a "*hopefull*" act (Harkness 2009, p. 6) and project of reconnection (Harkness 2009, p. 364); of production and consumption with their consequences; and of people with one another, their buildings, and their ecology.

Conclusion

This volume sought to explore the ways in which reuse practices can be seen to subvert consumerism. As Harkness (2009, p. 340) argues:

> Re-use discourses are generally 'alternative' and unofficial, and are anti-capitalist in that they contradict the consumer practice of repetitively buying new and throwing away 'old', a cycle which traditionally 'oils the wheels' of contemporary capitalism.

Such practices may seem relatively insignificant when viewed against the escalations and accelerations associated with the contemporary practices of mass consumption detailed in the opening chapters, and in many cases, could also be argued to be dependent on the very processes they seek to subvert. For example, in her chapter on ReNew Newcastle, Smith (this volume) describes how contributory economies are "simultaneously independent and interdependent, both resistant to, yet strangely reliant upon, the processes of mainstream capitalism" they seek to challenge. Such contradictions are characteristic of what Leahy (2015) refers to as the "hybrid" forms of late capitalism. A similar argument could be made about Earthships, as productively "parasitical formations" reliant on the "empty and abandoned shells of mainstream capitalism" but changing the system from within (or at least from the edge) (Smith, this volume, citing Braudel 1982, p. 373; Lindtner 2014, p. 145).

However, Earthships, permaculture, and the other 'reuse' movements discussed in this volume, cannot simply be dismissed as marginal movements, as their influence echoes far beyond their original locus. Paradoxically, it may well be that the 'marginal' nature of these movements is precisely what enables them to be so potentially transformative, as it enables them to influence mainstream trends while resisting total commodification and absorption by mainstream culture. In permaculture, this could be referred to as the "edge effect" where elements meet and provide opportunities for synergy. Earthship building and permaculture act "at the edge of chaos, at the edge of a change between systems", where information and knowledge from marginal movements "is echoed throughout the system" and "[s]mall changes can act as the trigger element to restructure a system" (Tippet 1994 cited in Birnbaum and Fox 2014, p. 14). As the permaculture principles suggest, "The edge is where it's at" (FVSS 2013 and Holmgren 2002).

Lockyer and Veteto (2013, p. 3) propose that anthropologists engage in "*ecocultural edgework* that moves beyond nature-culture dualisms and strengthens ongoing efforts to build a sustainable world." Earthship Ironbank is edgework in practice. As one of the first council-approved Earthships in Australia, the building has now set a precedent that is likely to influence the experience of many ownerbuilders seeking council approval for their own Earthship. Marty keenly shares his research and the knowledge he has accumulated through this design and approval process with those seeking to do the same and regularly opens his property for tours, open days and educational workshops about Earthships, permaculture, and 'natural' or 'eco' building. As a B&B, Earthship Ironbank will also provide its guests with information and first-hand experience of such topics, including what it is like to stay or 'dwell' in an Earthship. Unlike most of the 'dwelling' places discussed in Harkness's (2009) and Schelly's (2013, 2014, 2017) work, Earthship Ironbank was intended as a B&B, a place characterized by an influx of visitors in and out of the site, as has been the case with the workshops and site visits to date. There are numerous return visitors and people who have an ongoing relationship with the site. However, the nature of that relationship is less circumscribed and uncertain than current understandings of 'dwelling places' might suggest. While the long-term impacts of such a project of cultural transformation remain to be seen, as argued by Harkness (2009), the integrative design of Earthships challenges and transcends conventional boundaries between, for example, waste and non-waste, nature and culture, traditional and modern, production and consumption, work and play, expert and non-expert, buildings and people, buildings and their environment, and people and their environment. Earthships could thus be considered as 'edgework' 'par excellence'.

Drawing on ethnographic research conducted at Earthship Ironbank in the Adelaide Hills, this chapter has addressed the phenomenon of reuse in Earthship construction as a meaningful cultural practice affiliated with permaculture and 'natural', 'eco', or 'alternative' building movements. This chapter has also explored Earthship construction as a form of counter-hegemonic architecture, arguing that reuse in Earthship building challenges dominant paradigms of mass-production and hyper-consumption and presents renewed possibilities for re-imagining ways of living with the earth and with one another. Reuse and building with 'waste' in Earthship construction, leads not only to the construction of new physical structures, but also of emerging cultural and social systems, from the discarded materials of industrial society. As Miller (2010) argues, "Things make us just as much as we make things." An Earthship "both creates and grows forth from the relationships between people, between building and dweller, and between people and their environments" (Harkness 2009, p. 15). Earthship building can thus be understood as a form of "playbor" (Ritzer 2014b) that creatively employs strategies of design and reuse to reclaim and recreate the past to shape the future through transformative action in the present. However, Earthship building is just one of the many forms of reuse detailed in this volume. Such practices provide support for the argument that reuse can be employed as a mechanism to challenge and subvert dominant ideologies and practices associated with contemporary consumerism and present alternative approaches.

Acknowledgments

I would like to acknowledge and thank Dr. Martin (Marty) Freney for his pioneering work and Marty and his family for hosting me at Earthship Ironbank all these years. I would also like to thank all the workshop participants and 'Earthshippers' I have met over the years who shared their time and stories with me and allowed me to participate in this transformational collective experience.

Note

1 The grey water system was built but had to be decommissioned due to SA Health's concerns about safety. However, the system can be reinstated easily, and Freney hopes that future legislative changes will allow this.

References

Alaimo, S. (2010) *Bodily Natures: Science, Environment, and the Material Self*, Indiana University Press: Bloomington & Indianapolis.
Alexander, S. and Garrett, J. (2017) "The Moral and Ethical Weight of Voluntary Simplicity: A Philosophical Review", *Simplicity Institute Report 17a*, The Simplicity Institute: Melbourne, VIC, http://simplicitycollective.com/the-moral-and-ethical-weight-of-vol untary-simplicity-a-philosophical-review, accessed 21 January 2017.
Appelgren, S. and Bohlin, A. (2015) "Growing in Motion: The Circulation of Used Things on Second-hand Markets", *Culture Unbound: Journal of Current Cultural Research*, vol 7, pp. 143–168.
Australian Government (2014) "Product Stewardship for End-of-Life Tyres – Fact Sheet" Department of the Environment and Energy, www.environment.gov.au/protec tion/national-waste-policy/publications/factsheet-product-stewardship-end-life-tyres, accessed 21 January 2017.
BBC (British Broadcasting Company) (2009) *Grand Designs* "The Brittany Groundhouse: S09 E05" Channel 4, www.channel4.com/programmes/grand-designs/on-demand, accessed 21 January 2017.
BBC (2014) *Grand Designs* "The Brittany Groudhouse Revisited: S10 E09" Channel 4, www.channel4.com/programmes/grand-designs/on-demand and www.youtube.com/watch?v=qSIba-eKAbE, accessed 21 January 2017.
Bey, H. (2011) *T.A.Z. The Temporary Autonomous Zone*, United States: Pacific Publishing Studio.
Birnbaum, J. and Fox, L. (2014) *Sustainable Revolution: Permaculture in Ecovillages, Urban Farms, and Communities Worldwide*, Berkeley, CA: North Atlantic Books.
Block, E. (1986) *The Principle of Hope*, Cambridge, Mass: MIT Press.

Botsman, R. and Rogers, R. (2010) *The Rise of Collaborative Consumption*, New York: Harper Collins.

Boulding, K. E. (1966) "The Coming Economics of a Spaceship Earth", http://dieoff.org/page160.htm, accessed 21 January 2017.

Boym, S. (2001) *The Future of Nostalgia*, New York: Basic.

Brand, S. (1968–1998) The Whole Earth Catalog, www.wholeearth.com/index.php, accessed 6 August 2017.

Braudel, F. (1982[1979]). *Civilization and Capitalism 15-18th Century: Volume II: The Wheels of Commerce*, trans S. Reynolds. London: Collins.

Daly, K. (2014) "In Search of the Greater World: Journey Through the Mesa", *Permaculture South Australia: Biannual Journal of Permaculture SA*, Summer 2014/2015, pp. 19–22.

Dawdy, S. L. (2010) "Clockpunk Anthropology and the Ruins of Modernity", *Current Anthropology*, vol 51, no 6, pp. 761–793.

Douglas, M. (1966 [1994]) *Purity and Danger: An Analysis of the Concepts of Pollution and Taboo*, London; New York: Routledge.

Earthship Biotecture (2015a) "Earthship Design Principles", http://earthship.com/design-principles, accessed 21 January 2017.

Earthship Biotecture (2015b) "How to Make Bottle Bricks", http://earthship.com/blogs/2015/01/make-bottle-bricks/, accessed 22 January 2017.

Earthship Biotecture (2015c) "Urban Sky Autonomy Project", http:// http://earthship.com/new-york, accessed 25 July 2017.

Ecoflex Australia (2005) "The Environmental Impact of Tyre derived Products", www.ecoflexsoutheastnsw.com.au/casestudies/Ecoflex-Environmental-FAQs.pdf, accessed 21 January 2017.

Ecoshout (2015) "Earthship Biotcture: Learn to Build Self Sufficient Zero Waste Houses in New Mexico, USA with Earthship", www.ecoshout.org.au/jobs/learn-build-self-suffi cient-zero-waste-houses-new-mexico-usa-earthship, accessed 21 January 2017.

EPA (Environmental Protection Authority) 2016 "Container Deposits", www.epa.sa.gov.au/environmental_info/container_deposit, accessed 21 January 2017.

Ferguson, R. S. and Lovell, S. (2014) "Permaculture for Agroecology: Design, Movement, Practice, and Worldview: A Review", *Agronomy for Sustainable Development*, vol 34, no 2, pp. 251–274.

Freney, M. (2006) "Sustainable Strawbale", *ReNew Magazine*, no 94, January–March 2006, p. 38.

Freney, M. (2008) "Evolving Towards an Ecological Society", *Proceedings of the 3rd International Solar Cities Congress* 2008, 17–21 February 2008, Adelaide Convention Centre, South Australia.

Freney, M. (2009) "Earthships: Sustainable Housing Alternative", *International Journal of Sustainable Design*, vol 1, no 2, pp. 223–240.

Freney, M. (2014) "Earthship Architecture: Post-occupancy Evaluation, Thermal Performance and Life-cycle Assessment", PhD Thesis, The University of Adelaide, https://digital.library.adelaide.edu.au/dspace/handle/2440/91872.

Freney, M. (2015) "From Earth, Cans and Tyres: Earthship Ironbank", *Renew Magazine*, October–December, vol 133, pp. 56–60.

Freney, M., Soebarto, V. and Williamson, T. J. (2012) "Learning from 'Earthship' Based on Monitoring and Thermal Simulation", Paper presented at the 46th Annual Conference of the Architectural Science Association, Griffith University, Queensland Australia.

Freney, M., Soebarto, V. and Williamson, T. J. (2013a) "Earthship Monitoring and Thermal Simulation", *Architectural Science Review*, vol 56, no 3, pp. 208–219.

Freney, M., Soebarto, V. and Williamson, T. J. (2013b) "Thermal Comfort of Global Model Earthship in Various European Climates", Paper presented at the 13th International conference of the International Building Performance Simulation Association, Chambery, France.

Fuller, B. (1969 [2014]) *Operating Manual for Spaceship Earth*, Zurich: Lars Muller Publishers.

FVSS (Formidable Vegetable Sound System) (2013) *Permaculture: A Rhymer's Manual*, Australia: Grow Do It.

Grasseni, C. (2007) *Skilled Visions: Between Apprenticeship and Standards*, Oxford; New York: Berghahn Books.

Haraway, D. (1997) *Modest Witness@ Second Millennium FemaleMan Meets OncoMouse: Feminism and Technoscience*, New York: Routledge.

Harkness, R. J. (2009) *Thinking Building Dwelling: Examining Earthships in Taos and Fife*, PhD Thesis, University of Aberdeen.

Harkness, R. J. (2011) "Earthships: The Home that Trash Built", *Anthropology Today*, vol 3, no 1, pp. 54–65.

Hawkins, G. and Muecke, S. (2003) *Culture and Waste: The Creation and Destruction of Value*, Maryland: Rowman and Littlefield Publishers Inc.

Helbig, K. (2017a) "Natural Energy", *SA Life Magazine*, February, pp. 88–92.

Helbig, K. (2017b) 'Earthship Ironbank', *PIP Magazine*, no 8, p. 44.

Hewitt, M. and Telfer, K. (2012) *Earthships in Europe* (2nd ed.), Bracknell: IHS BRE Press.

Hickson, P. (2015) "Earth-Building – How Does it Rate?", in D. Ciancio and C. Beckett (eds) *Rammed Earth Construction: Cutting Edge Research on Tradition and Modern Rammed Earth*, London: Taylor and Francis.

Hodge, O. (2009) *Garbage Warrior*, video, Australia: Hopscotch.

Holmgren, D. (2002 [2011]) *Permaculture: Principles & Pathways Beyond Sustainability*, Hepburn, Australia: Holmgren Design Services.

Holmgren, D. (2009) *Future Scenarios: How Communities Can Adapt to Peak Oil and Climate Change,* White River Junction, VT: Chelsea Green Publishing, www.futurescenarios.org/, accessed 23 January 2017.

Holmgren, D. (2011) *Sustainable Living Festival: Introduction to Michael Reynolds*, Hepburn, Australia: Holmgren Design Services.

Holmgren, D. (2013) *Essence of Permaculture*, Holmgren Design Services: Hepburn, VIC, http://holmgren.com.au/downloads/Essence_of_Pc_EN.pdf, accessed 21 January 2017.

Holmgren, D. (2017) "Permaculture Design Frameworks, Ethics and Principles", CQUniversity Permaculture Graduate Certificate Presentation, April 2017, Adelaide.

Humboldt State University (n.d.) "Do Earthship Tire Constructions Pose Health Risks?", Campus Center for Appropriate Technology (CCAT), www.ccathsu.com/askccat/do-earthship-tire-constructions-pose-health-risks/, accessed 21 January 2017.

Husband, K. (2013) "Earthship Is Out of This World", *The Adelaidean*, Spring, www.adelaide.edu.au/adelaidean/issues/64582/news64663.html, accessed 21 January 2017.

Illich, I. (1973) *Tools for Conviviality*, New York: Harper and Row.

Leahy, T. (2015) "Permaculture Practice: Hybrids of the Gift Economy and Capitalism", *The 12th Australasian Permaculture Convergence*, Penguin, Tasmania, 9–12 March, http://gifteconomy.org.au/the-gift-economy/permaculture-practice-hybrids-of-the-gift-economy-and-capitalism/, accessed 6 February 2017.

Leopold, A. (1949) *A Sand County Almanac with Essays and Conversations from Round River*, New York: Random House.

Levi-Strauss, C. (1962) *The Savage Mind*, Oxford: Oxford University Press.

Lindtner, S. (2014) "Hackerspaces and the Internet of Things in China: How Makers Are Reinventing Industrial Production, Innovation, and the Self", *China Information*, vol 28, no 20, pp. 145–167.

Lipovetsky, G. (2011) "The Hyperconsumption Society", in K. M. Ekström and B. Glans (eds) *Beyond the Consumption Bubble*, London: Routledge.

Lockyer, J. and Veteto, J. R. (2013) *Environmental Anthropology Engaging Ecotopia: Bioregionalism, Permaculture, and Ecovillages*, New York: Berghahn Books.

Lozano, D. V. (2010) *Earthships: New Solutions*, London: Detachment East.

Luckman, S. (2013) "The Aura of the Analogue in a Digital Age: Women's Crafts, Creative Markets and Home-Based Labour after Etsy", *Cultural Studies Review*, vol 19, no 1, pp. 249–270.

Luckman, S. (2015) *Craft and the Creative Economy*, London: Palgrave Macmillan.

Macy, J. and Johnstone, C. (2012) *Active Hope: How to Face the Mess We're in Without Going Crazy*, Novato, CA: New World Library.

Maniates. M. and Meyer, J. M. (eds) (2010) *The Environmental Politics of Sacrifice*, Cambridge, MA: MIT Press.

Marx, K. (1977) "Thesis on Fuerbach", in D. McLellan (ed) *Kark Marx Selected Writings*, Oxford: Oxford University Press.

McCloud, K. (2015) "Episode Three: Belize", *Kevin McCloud's Escape to the Wild*, Channel 4, UK.

Merrifield, A. (2000) 'Henri Lefebvre: A Socialist in Space', in M. Crang and N.J. Thrift (eds.) *Thinking Space*. New York: Routledge.

Miller, D. (2010) *Stuff*, Cambridge: Polity Press.

Mollison, B. (1979) *Perma-Culture Two: Practical Design for Town and Country in Permanent Agriculture*, Tasmania: Tagari Publications.

Mollison, B. and Holmgren, D. (1978) *Perma-Culture One: A Perennial Agriculture for Human Settlements*, Tasmania: Tagari Publications.

Mollison, B. (1988) *Permaculture: A Designer's Manual*, Tasmania: Tagari Publications.

Open Eye Media (2015) *Garbage Warrior*, Australia: Hopscotch, www.garbagewarrior.com/about, accessed 21 January 2017.

Osmond, J. and Alexander, S. (2016) *A Simpler Way: Crisis as Opportunity*, Happen Films, www.youtube.com/watch?v=XUwLAvfBCzw, accessed 1 January 2017.

Preston Prinz, R. (2015) *Hacking the Earthship: In Search of an Earth-Shelter that Works for Everybody*, Albuquerque, New Mexico: Archinia Press.

Price, A. (2009) *Slow-Tech: Manifesto for an Over-Wound World*, London: Atlantic Books.

Reynolds, M. (1985) *Water From the Sky*. Taos, New Mexico: Solar Survival Press.

Reynolds, M. (1989) *A Coming of Wizards: A Manual of Human Potential*, Taos, New Mexico: The High Mesa Foundation.

Reynolds, M. (1990) *Earthship Volume 1: How to Build Your Own*, Taos, New Mexico: Solar Survival Press.

Reynolds, M. (2008) *Journey Part 1*, Taos, New Mexico: Earthship Biotecture.

Reynolds, M. (2018) "Earthship Design Principles", https://www.earthshipglobal.com/design-principles/, accessed 12 April 2018.

Ritzer, G. (2010) "Production, Consumption, Prosumption", *Journal of Consumer Culture*, vol 10, no 1, pp. 13–36.

Ritzer, G. (2014a) "Prosumption: Evolution, Revolution, or Eternal Return of the Same?" *Journal of Consumer Culture*, vol 14, no 1, pp. 3–24.

Ritzer, G. (2014b) "E-Games and Prosumption", 10 September, https://georgeritzer.wordpress.com/category/prosumption/, accessed 21 January 2017.

Ritzer, G., Dean, P. and Jurgenson, N. (2012) "The Coming of Age of the Prosumer", *American Behavioral Scientist*, vol 56, no 4, pp. 379–398.

Rosa, H. (2003) "Social Acceleration and the Political Consequences of a Desynchronised High-Speed Society", *Constellations*, vol 10, no 1, pp. 3–33.

Scanlan, J. (2005) *On Garbage*, London: Reaktion Books.

Schelly, C. (2013) *Dwelling in Resistance: Living with Alterative Technologies in America*, PhD Thesis, The University of Wisconsin.

Schelly, C. (2014) "Are Residential Dwellers Marking and Claiming? Applying the Concepts to Humans Who Dwell Differently", *Environment and Planning D: Society and Space*, vol 32, pp. 672–688.

Schelly, C. (2017) "Self-Sufficiency as Social Justice: The Case of Earthship Biotecture", in C. Schelly (ed) *Dwelling in Resistance Living with Alternative Technologies in America*, Rutgers University Press (forthcoming – chapter viewed ahead of print).

Seriff, S. (1996) "Folk Art from the Global Scrap Heap: The Place of Irony in the Politics of Poverty", in C. Cerny and S. Serfiff (eds) *Recycled, Re-seen: Folk-Art from the Global Scrap Heap*, New York: Haryy N. Abrams, pp. 8–29.

Sharpe, T. (2000) "Controversy Over Green Hero", *Architectural Record*, vol 188, no 6, p. 36.

Sheen, D. (2009) *First Earth: Uncompromising Ecological Architecture*, www.davidsheen.com/firstearth/film.htm, accessed 21 January 2017.

Sklair, L. (2010) "Iconic Architecture and the Culture-ideology of Consumerism", *Theory of Culture and Society*, vol 27, no 5, pp. 135–159.

Sklair, L. and Gherardi (2012) "Iconic Architecture as the Hegemonic Project of the Transnational Capitalist Class", *City*, vol 16, no 1–2, pp. 57–73.

Sterbenz, C. (2014) "How an Environmentalist Architect Made an Incredible House Out of Garbage", *Business Insider*, 25 July, www.businessinsider.com/michael-reynolds-beer-can-thumb-house-2014-7?op=1, accessed 21 January 2017.

Stockmann, E. (2017) "An Earthship Village", *Owner Builder Magazine*, vol 199, February/March, pp. 71–75.

Smith, C. (in press) "Permaculture – History and Futures", in R. Slaughter and A. Sykes-Kelleher (eds), *History of Australian Futures*, Centre for Australian, http://cfaf.com.au/a-history-of-australian-utures/, accessed 23 January 2017.

Telfer, K. (2003) 'Earth Mover', *The Architects Journal*, vol 19, pp. 18–19.

Tinniswood, A. (1999) *The Arts and Crafts House*, London: Mitchell Beazley.

Tippet, J. (1994) "A Pattern Language of Sustainability: Ecological Design and Permaculture", B.A. Dissertation, Lancaster University.

Today Tonight (2015) "Sustainable Housing", 21 September 2015, www.todaytonightadelaide.com.au/stories/sustainable-housing, accessed 21 January 2017.

Toffler, A. (1990) *The Third Wave*. London: COllins.

Totally Wild (2015) "Season 23 Episode 57", 21 November (from 15min 12sec), 21 November 2015, http://tenplay.com.au/channel-eleven/totally-wild/season-23/episode-57, accessed 21 January 2017.

Turner, V. (1969) *The Ritual Process: Structure and Anti-structure*, New York: Aldine De Gruyter.

TV3 (2015) "Series 1 Ep 2 – Coromandel", *Grand Design New Zealand*, TV3: New Zealand.

Urry, J. (2007) *Mobilities*, New York: Wiley.

Van Gennep, A. (1960) *The Rites of Passage*, London: Routledge and Kegan Paul.

Wilcock, R. (2016) 'Evaluating the Earthship – a Effective Sustainable Building Concept or Not'. *Interesting Engineering*, 9th Feb 2016. https://interestingengineering.com/evaluating-the-earthship-an-effective-sustainable-building-concept-or-not. Accessed 26th May 2018.

Zimring, C. A. and Rathje, W. L. (eds) (2012) *Encyclopedia of Consumption and Waste: The Social Science of Garbage, Volume 1*, London: Sage.

Index

Note: Page numbers in *italics* indicate figures and page numbers in **bold** indicate tables on the corresponding pages.